Principles and Applications of Data Science

Principles and Applications of Data Science

Editor

Chuan-Ming Liu

MDPI • Basel • Beijing • Wuhan • Barcelona • Belgrade • Manchester • Tokyo • Cluj • Tianjin

Editor
Chuan-Ming Liu
National Taipei University of Technology (Taipei Tech)
Taiwan

Editorial Office
MDPI
St. Alban-Anlage 66
4052 Basel, Switzerland

This is a reprint of articles from the Special Issue published online in the open access journal *Applied Sciences* (ISSN 2076-3417) (available at: https://www.mdpi.com/journal/applsci/special_issues/data_sciences).

For citation purposes, cite each article independently as indicated on the article page online and as indicated below:

LastName, A.A.; LastName, B.B.; LastName, C.C. Article Title. *Journal Name* **Year**, *Volume Number*, Page Range.

ISBN 978-3-0365-4527-1 (Hbk)
ISBN 978-3-0365-4528-8 (PDF)

© 2022 by the authors. Articles in this book are Open Access and distributed under the Creative Commons Attribution (CC BY) license, which allows users to download, copy and build upon published articles, as long as the author and publisher are properly credited, which ensures maximum dissemination and a wider impact of our publications.
The book as a whole is distributed by MDPI under the terms and conditions of the Creative Commons license CC BY-NC-ND.

Contents

About the Editor

Chuan-Ming Liu

Dr. Chuan-Ming Liu is a professor at the Department of Computer Science and Information Engineering (CSIE), National Taipei University of Technology (Taipei Tech), Taiwan, where he was the Department Chair from 2013 to 2017 as well as the Head of the Extension Education Center at the same school from 2018 to 2021. Dr. Liu received his Ph.D. in Computer Science from Purdue University in 2002 and joined the CSIE Department in Taipei Tech in the spring of 2003. In 2010 and 2011, he held visiting appointments with Auburn University, Auburn, AL, USA, and the Beijing Institute of Technology, Beijing, China. He has a role in many journals, conferences, and societies and has published more than 100 papers in many prestigious journals and international conferences. Dr. Liu was the co-recipient of many best paper awards, including ICUFN 2015, ICS 2016, MC 2017, WOCC 2018, MC 2019, MC 2021, and WOCC 2021. His current research interests include big data management and processing, uncertain data management, data science, spatial data processing, data streams, ad-hoc and sensor networks, and location-based services.

MDPI

Article

Predicting Implicit User Preferences with Multimodal Feature Fusion for Similar User Recommendation in Social Media

Jenq-Haur Wang [1,*], Yen-Tsang Wu [1] and Long Wang [2,*]

[1] Department of Computer Science and Information Engineering, National Taipei University of Technology, Taipei 106344, Taiwan; buddyswu@gmail.com

[2] School of Computer and Communication Engineering, University of Science and Technology Beijing, Beijing 100083, China

* Correspondence: jhwang@csie.ntut.edu.tw (J.-H.W.); lwang@ustb.edu.cn (L.W.)

Abstract: In social networks, users can easily share information and express their opinions. Given the huge amount of data posted by many users, it is difficult to search for relevant information. In addition to individual posts, it would be useful if we can recommend groups of people with similar interests. Past studies on user preference learning focused on single-modal features such as review contents or demographic information of users. However, such information is usually not easy to obtain in most social media without explicit user feedback. In this paper, we propose a multimodal feature fusion approach to implicit user preference prediction which combines text and image features from user posts for recommending similar users in social media. First, we use the convolutional neural network (CNN) and TextCNN models to extract image and text features, respectively. Then, these features are combined using early and late fusion methods as a representation of user preferences. Lastly, a list of users with the most similar preferences are recommended. The experimental results on real-world Instagram data show that the best performance can be achieved when we apply late fusion of individual classification results for images and texts, with the best average top-k accuracy of 0.491. This validates the effectiveness of utilizing deep learning methods for fusing multimodal features to represent social user preferences. Further investigation is needed to verify the performance in different types of social media.

Keywords: deep learning; user preference learning; feature fusion; similar user recommendation

Citation: Wang, J.-H.; Wu, Y.-T.; Wang, L. Predicting Implicit User Preferences with Multimodal Feature Fusion for Similar User Recommendation in Social Media. *Appl. Sci.* **2021**, *11*, 1064. https://doi.org/10.3390/app11031064

Received: 23 December 2020
Accepted: 21 January 2021
Published: 25 January 2021

Publisher's Note: MDPI stays neutral with regard to jurisdictional claims in published maps and institutional affiliations.

Copyright: © 2021 by the authors. Licensee MDPI, Basel, Switzerland. This article is an open access article distributed under the terms and conditions of the Creative Commons Attribution (CC BY) license (https://creativecommons.org/licenses/by/4.0/).

1. Introduction

In popular online social networks such as Twitter, Facebook, and Instagram, it is easy for users to share information and post their opinions and comments. Given the huge amount of user-generated content (UGC), it is difficult to search for the most relevant information effectively. Since people tend to share information that interests them and comment on the topics they like, user posts and comments are likely to reflect their preferences. In addition to individual posts, it would be useful if we can also recommend groups of people with similar interests. This might help users discover more relevant content in social media. Conventionally, characteristics of users are often defined by their demographic information, such as gender, age group, occupation, and income status. These are usually obtained by filling in surveys and basic personal information which is often lacking since users are reluctant to provide this in most social media. Thus, it is a challenge to learn detailed user preferences without explicit user feedback.

Conventional recommender systems focus on suggestion of items as a function of similar items or users in previous records. However, user behaviors in online social networks are different from e-commerce websites since there is no explicit feedback from users regarding the relevance assessment of items. Instead, users participate in social media by posting, replying (or retweeting), and forwarding multimedia content. Since users might post various types of multimedia content related to the corresponding texts in the surrounding paragraphs, it would be useful if these contextual relations could be utilized to discover implicit user preferences. In this paper, we propose a multimodal feature fusion approach to implicit user preference prediction by learning deep features from texts and images in user posts for recommending similar users in social media. First, text and image features in user posts are extracted using deep learning techniques, such as convolutional neural networks (CNNs) and TextCNN. Then, these features are combined using multimodal feature fusion methods as a potential representation of user preferences. Lastly, similarity among user preferences can be defined and calculated for finding and recommending similar users. Since there are multiple types of features, in this paper, we compare the effects of early and late fusions of features for user preference prediction. To evaluate the performance of user recommendation, a dimensionality reduction method using autoencoders is compared with user clustering.

From the experimental results of Instagram data, we can see the clear advantage of deep learning and feature fusion over individual features. Recommending the top-k similar users when applying late fusion of individual classification results of texts and images gives the best average top-k accuracy of 0.491. This shows the potential of discovering implicit user preferences from multimodal content posted by users. Further investigation is needed to verify the effects in different types of social media.

The major contributions of this paper can be summarized as follows:

- First, we propose a convolutional deep learning method for extracting image and text features from user posts as a potential representation of implicit user preferences.
- Second, we compare feature fusion methods to combine text and image features from user posts for predicting user preferences.
- In our experimental results on real-world Instagram data, the best average top-k accuracy of 0.491 for recommending top-50 similar users can be obtained when applying late fusion on text and image features. This shows an improvement of 36.3% over the baseline in terms of accuracy.

2. Related Works

There are two categories of studies related to our work. One involves techniques that model user preferences by items and user's information for user recommendation. The other involves deep learning techniques adopted for feature extraction and fusion.

For user or people recommendation in social networks, most existing approaches rely on social relations and network structures in addition to content similarity. For example, Chen et al. [1] found out that the social network structure tends to give known contacts, while content similarity helps to find new friends. Hannon et al. [2] considered content-based techniques and collaborative filtering approaches based on followees and followers of users.

Armentano et al. [3] proposed to recommend relevant users by exploring the topology of the network. Since content-based approaches tend to have low precision while collaborative filtering based approaches based on follower-followee relations have data sparsity issues, Zhao et al. [4] proposed a community-based approach that utilizes an Latent Dirichlet Allocation (LDA)-based method on follower–followee relations to discover communities before applying matrix factorization for user recommendation. Gurini et al. [5] proposed to extract semantic attitudes from user-generated content, including sentiment, volume, and objectivity, and they conducted people recommendation using matrix factorization. In this paper, since there are no social network structures available, we used content-based approach as our baseline model. Moreover, we included word embedding using a pretrained Word2Vec model to get semantic information of texts.

Predicting user preferences is very important when constructing a recommender system. In social networks, user preferences can be derived from three sources. The first is user post contents including texts and images. They provide the direct evidence of what users like. The second is user interested topics, which could be reflected from tags in posts. The third is the user relations in social networks. Recent preference prediction models integrate information including user posts [6], images [7,8], social network attributes [9], and user demographic information [10]. Examples include gender, age, and political tendency [11]. In past research, researchers considered integrating information such as reviews and social relations to predict user preferences. For example, some methods learned user sentiments from user reviews [12,13] and item topics regarding user preferences [14,15]. Some methods used hybrid methods to learn user's opinion in different domains [8,16]. Unlike the above studies, we did not use the extra information such as social relations or user demographics. Instead, we focused on extracting information from post contents, including texts and images.

On the other hand, deep learning practices have been applied to texts to provide more insights into the reasons behind users' preferences and more awareness of item features they consider relevant [17]. Deep learning has been shown to be effective in user preference learning [18,19]. Palangi et al. [20] proposed a deep model which used a variant of recurrent neural network (RNN) architecture called Long Short-Term Memory (LSTM) for retrieval task. The study by Tai [21] used Tree-LSTM to predict semantic connection of sentences and sentiment classification. Yousif et al. [22] combined CNN and Bidirectional LSTM (Bi-LSTM) to analyze citation sentiment and purpose classification. Seo et al. [23] used CNN with attention to model user preferences and item properties as expressed in review texts. However, these methods take more computation due to their complex structure. In contrast, our proposed method focuses on fusing different features instead of relying on complex models to achieve good performance in classification.

Some existing preference prediction methods utilized users' review texts to learn user preference. In Chambua et al. [24], they used the hybrid approach to learn and represent user preferences and predict them by using RNN–LSTM and probabilistic matrix factorization in the Amazon Products Datasets. Because they only used review texts as the user feature, they are faced with the problem of missing data. In Lv et al. [25], the visual and social features were fused by linear regression, matrix factorization, and support vector regression. Unlike our proposed method, they used tags and titles as the textual feature. This might not accurately capture the user's true emotional preferences. Zhang et al. [26] used the attention mechanisms to extract textual and visual features by a variant of CNN called VGGnet and LSTM in the Flickr dataset. Through a linear attention mechanism, they fused the textual, visual, and user features for prediction. With the same problems as Lv et al. [25], the textual description of posts only included titles and tags which might not reflect emotional preferences. Aloufi et al. [27] used the visual and social features and information-associated content to predict popular images by ranking the Support Vector Machine (SVM) model in the Flickr dataset. They added several extra features to predict which photos would be popular. Mazloom et al. [28] used the Instagram dataset for their experiment. They used features from users, items, and contexts of posts as representation to predict the popularity of a post related to a specific user and item in social media by matrix factorization. The difference to our proposed method is that they added the visual and textual sentiment based on Visual Sentiment Ontology [29] and SentiStrength [30]. Unlike the above papers, we represent user preferences by fusing text contents and image features extracted with convolutional neural networks. Furthermore, we address the issue of user recommendation by autoencoders and user clustering.

Feature fusion is an important method in pattern recognition which allows for more robust predictions by incorporating multiple features that might complement each other. When some of the features are missing, we can still make predictions. Contextual similarity [31] has been extensively exploited recently in retrieval tasks, such as biological information retrieval, natural image search, shape retrieval, and analysis of time series. A more recent example is the unsupervised ranking model in which all words of a query or document are embedded into vectors, which are matched by deep neural networks [32]. The multimodal approach was constructed by different input sources [33,34]. For example, the multimodal approach based on image and text features was employed in multiple tasks such as retrieval, classification, and natural language processing. To find relationships between text and image features, two general multimodal fusion approaches were deployed: early and late fusions [35]. Features with poor performance will greatly affect the effects of early fusion [36]. In contrast to early fusion, late fusion uses mechanisms such as averaging [37], voting [38], and learned model [39,40] to fuse predictions from each model. In our paper, we aimed to compare the effects of early and late fusions in user preference prediction.

3. The Proposed Method

In this paper, we propose a feature fusion approach to implicit user preference learning from user posts and related images for similar user recommendation. The overall system architecture is illustrated in Figure 1.

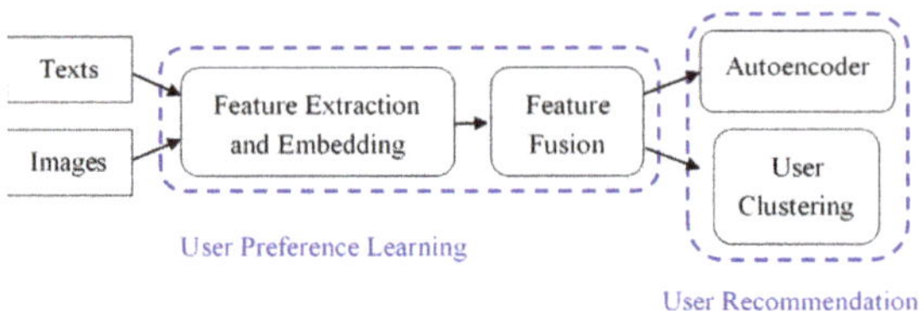

Figure 1. The system architecture of our proposed method.

As shown in Figure 1, there are three modules in the proposed method: feature extraction and embedding, feature fusion, and user recommendation. First, user posts in social networks are collected as the dataset, where texts and images are extracted by using text and image convolutional networks, respectively. Second, text and image features are combined using feature fusion techniques as the representation of user preferences. Lastly, on the basis of the similarity among user preferences, top similar users are recommended by autoencoders and user clustering.

3.1. Feature Extraction

People often express their opinions on selected topics by posting related texts and images they are interested in. These behaviors might show their implicit user preferences. Thus, to better understand what people like, in this paper, we assume that user preferences can be represented by the characteristics of texts and images in their posts and comments.

Previous studies used the title or tags as text features for classification. However, user preferences cannot be effectively learned since tags might be ambiguous and diverse in their meanings. Since text contents usually contain more semantic information such as emotions or stances than tags, text contents are used instead of tags. In this paper, we focus on deep learning methods such as the TextCNN model [41] for text feature extraction. First, text documents are represented by a word embedding model such as Word2Vec [42,43]. This is a distributional representation of words among different contexts in fix-sized vectors. Next, we use TextCNN [41] to extract text features. This is a slight variant of the CNN architecture by Collobert et al. [44]. The TextCNN architecture is shown in Figure 2.

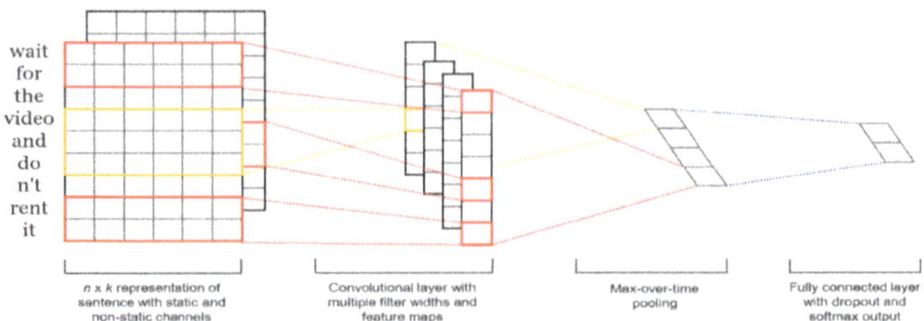

Figure 2. Convolutional neural networks (CNNs) for text classification [41].

As shown in Figure 2, there are four layers in the TextCNN model: word embedding layer, convolutional layer, max pooling layer, and fully connected layer. First, given a sentence of length n, we denote it by its word embeddings as follows:

$$x_{1:n} = x_1 \varnothing x_2 \varnothing \ldots \varnothing x_n, \tag{1}$$

where x_i is the embedding of word i with dimension k, and $\varnothing$ is the concatenation operator. Second, in the convolutional layer, the convolution operation involves a filter $w \in R^{hk}$, which is applied to a window of h words to produce a feature c_i for the i-th window as follows:

$$c_i = f(w * x_{i:i+h-1} + b), \tag{2}$$

where $b \in R$ is a bias term and f is a nonlinear function, such as the hyperbolic tangent. For the given sentence $x_{1:n}$, this filter produces a feature map c as follows:

$$c = [c_1, c_2, \ldots, c_{n-h+1}], \tag{3}$$

where $c \in R^{n-h+1}$. Then, in the max pooling layer, we take the maximum value $c' = max(c_1, c_2, \ldots, c_{n-h+1})$ as the feature map corresponding to this particular filter. Lastly, the fully connected layer outputs the final classification result or category. In this paper, we also extract outputs from the layer before the last layer to compare the effects of feature fusion.

To extract image features, we utilize a CNN model called VGG16 [45] to convolve the image pixels represented in the Red-Green-Blue (RGB) color model. This architecture contains several differences from the previous convolutional networks. It is a multilayer convolutional network that is a thorough evaluation of networks of increasing depth using an architecture with very small (3 × 3) convolution filters, which shows a significant improvement by increasing the depth to 16 layers (13 convolutional layers and three fully connected layers). Layers of configurations are designed using the same principles, inspired by Ciresan et al. [46] and Krizhevsky et al. [47]. Compared with other models such as [48], VGG16 is a CNN architecture with fewer parameters but comparable accuracy. For this reason, it is used as the Image CNN model to extract the feature map from images.

3.2. Feature Fusion

With texts and images in a user post, we try to combine them in two different ways, i.e., early fusion and late fusion.

3.2.1. Early Fusion

In early fusion, embeddings of texts and images are input to the same CNN model simultaneously for feature extraction. Early fusion is also known as feature-level fusion, which can be expressed as follows: $x_m = f(x_1, \ldots x_n)$, where an aggregated representation x_m of features is computed by function f that integrates individual features $x_1, \ldots, x_n$. Early fusion combines different input sources into a single feature vector, which is used as inputs to the classification framework. The advantage of early fusion is that it learns all the features in one phase. This makes the training pipeline easier, but a lot of important information might be lost. The architecture is shown in Figure 3.

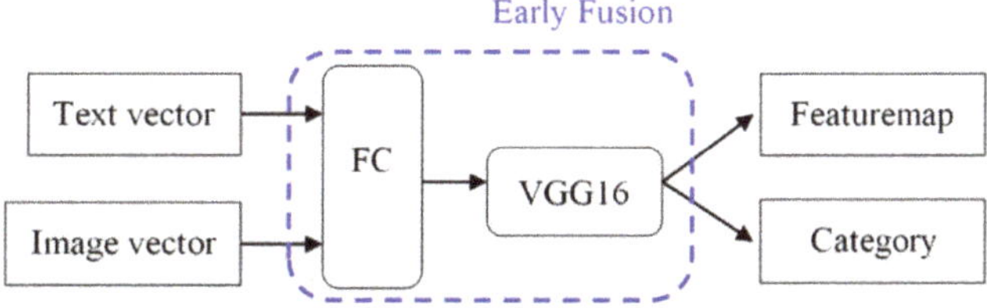

Figure 3. The architecture for early fusion of text and image features.

As shown in Figure 3, first, since the sizes of text and image embeddings are different, we use a fully connected (FC) neural network to fuse these vectors. There are two different ways of concatenation: text vector concatenated with images or image vector concatenated with texts. Next, we extract the output of different layers in the VGG16 model including the feature map and the final classification result or category as the combined representation for the texts and images.

3.2.2. Late Fusion

In late fusion, text and image features are extracted and then combined for classification. This is also known as decision-level fusion [49]. It integrates different model predictors by a fusion mechanism to come up with the final decision. Late fusion can be expressed as follows: *output* = g(f_1 (x_1), … ,f_n(x_n)), where functions f_1, … ,f_n are applied to individual features and function g is used to aggregate the individual decisions by f_1, … ,f_n. The main disadvantage of late fusion is that it cannot learn the correlation among features. Compared to early fusion, late fusion tends to be more robust to features that have negative influence [36]. The architecture is shown in Figure 4.

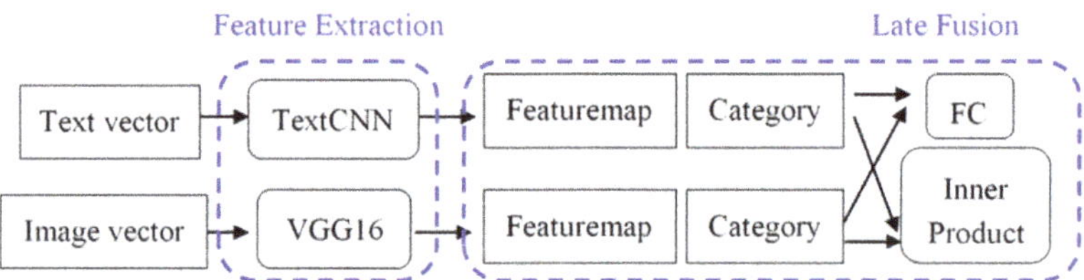

Figure 4. The architecture for late fusion of text and image features.

As shown in Figure 4, after embedding texts and images into vectors by Word2Vec and RGB models, respectively, they are input to two separate CNN models for feature extraction as described in Section 3.1. The corresponding feature map and text/image category can be obtained. Then, to fuse text and image features, in addition to using a fully connected layer, we also try to combine the two feature maps by inner product since they are of the same size. The idea is to increase the correlation and reduce the dimension through the inner product.

3.3. *User Preference Learning*

After combining text and image features using either early or late fusion techniques, we obtain the corresponding feature map and category for each user post. To further represent user preferences, there are two different methods. First, since feature map is the internal representation of a post, the centroid of the feature maps of all posts can be regarded as the user preferences. We can simply add all features with the addition operator to get the centroid. Second, the category of each user post is assumed to reflect part of the user preferences since people tend to post topics they are interested in. People with the same preferences are more likely to post in the same category. Therefore, for each user, we count the number of posts in each category and find out the majority category as the user preference. These are shown in Equations (4) and (5).

$$f_c = F_a([c_1, c_2, \ldots, c_n]), \tag{4}$$

$$f_R = F_c([R_1, R_2, \ldots, R_n]), \tag{5}$$

where F_a is the addition operator and F_c is the count operator, whereas c_i and R_i are the feature maps and categories of the post, respectively. Since the output categories R_i in neural networks are usually represented by one-hot encoding, we can simply accumulate them using the count operator to find out the majority category. By aggregating the feature map and category of all images and texts posted by the user, we can obtain the potential feature for user preferences.

3.4. Dimension Reduction

From preliminary observations, we found that feature maps in images and texts usually have very high dimensions, which could be as high as 8192 after feature fusion. In addition to computational time complexity, it may also include a lot of noise inside. To find out the most important features, we further apply autoencoders for dimension reduction.

Given a fixed dimensional representation of the user preferences as inputs, we design the autoencoder as a single-hidden-layer d-dimensional neural network. Autoencoders consist of an encoder that transforms input to a code, and a decoder which reconstructs the input from the code. When the number of neurons in the hidden layer of autoencoders is less than that in the input layer, it is forced to learn the compressed representation of the input data. After dimensionality reduction, our model performs similarity calculations among users to generate a list of recommendations. The architecture is shown in Figure 5.

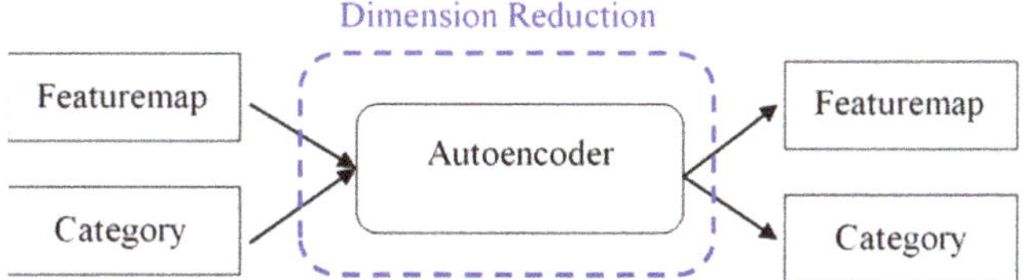

Figure 5. The user preference learning architecture with dimension reduction.

Since the goal of using autoencoders is to reduce dimensionality by simple neural networks, it would be beneficial if the dimensionality can be reduced while maintaining comparable performance in recommendation. In order to find out the most compact representation which is still effective in recommendation, we conducted experiments to verify the effects of different dimensionality d on the performance of user recommendation.

3.5. Recommendation Based on User Clustering

When there are many users, it takes a lot of time to calculate similarity between each one according to the aforementioned method. It is hard to give real-time recommendation. Thus, we further investigate the effect of clustering all users by the K-means algorithm to reduce the number of candidates for similar user recommendation.

Although K-means clustering is incremental, to save time when updating clusters, we propose to divide it into two stages: offline clustering and online recommendation. First, users are represented by the output category after learning user preferences in Section 3.3 and then clustered using the K-means algorithm. Since the dimension of the category vector is lower than that of feature maps, it reduces the computational time. Then, we calculate cosine similarity by using the feature map of each member in the same cluster to find out the most similar members to recommend. Because the number of users in each group is much smaller than all users, it also reduces the computational time for online recommendation if user clustering can be done offline.

If the original user posts a new post, we only need to calculate the similarity for the members of the same group. If we have a new member, we just need to calculate which cluster it belongs to for recommendation and then calculate the similarity between cluster members, instead of all users. In addition, since clustering can be done offline, the method greatly improves the efficiency of online recommendation. The architecture is shown in Figure 6.

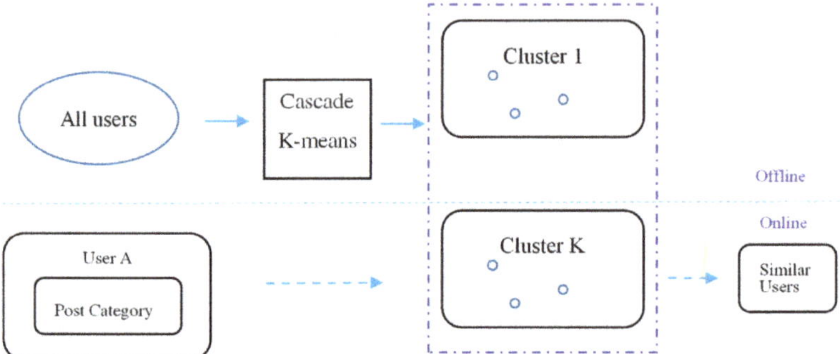

Figure 6. The architecture of recommendation based on user clustering.

The problem for conventional K-means algorithm is to determine the best number of clusters. Since Cascade K-means [50,51] can automatically determine the optimal number of clusters, it was selected to solve this problem. We use the classification result of the user as inputs for user clustering with Cascade K-means. Lastly, after calculating cosine similarity among the members in the same cluster, we can obtain the most similar K members to recommend.

Although we did not deploy it as a potential real-world recommendation system, we could implement different types of thresholds before initiating the update process by clustering. For example, we could count the number of new users, and set the threshold for a new user ratio. Furthermore, when the percentage of new posts by a user reaches a threshold, we could initiate an update process to update the cluster members for more accurate recommendation.

4. Experiments

In this section, we describe our datasets, how training and testing are performed, and our baseline algorithms, and we analyze the experimental results.

To evaluate the performance of the proposed method, we used the current mainstream social media—Instagram—in our experiments. To the best of our knowledge, there is no publicly available social media dataset that includes both images and texts posted by the same user. To collect the Instagram dataset, we randomly selected six hashtags from the list of top 100 hashtags on Instagram in 2018 as queries to get user posts. This can be done using the Hashtag Search API, which is only available for Instagram Professional accounts. Thus, we had to directly crawl the latest updates for each user on Instagram webpages for tag-based queries. There is a limit of a maximum number of 45 posts per user that can be crawled. To avoid large variations in user participation where there were too few posts to learn the features, we chose to keep only users who had posted at least nine posts, which is a 5:1 ratio (20% of the maximum). Each post contained the user identifier (ID) and contents of the user posts and hashtags. This dataset was used to train the classification models of user preferences (denoted as Dataset_1). There were 239 users and 6941 posts in Dataset_1, from which we split into 5553 posts for training and 1388 posts for testing. The major language was English.

Considering the diversity in linguistic form, we used two pretrained Word2Vec embeddings as dictionaries for English and Chinese. The detailed attributes of the word embeddings are outlined in Table 1.

Table 1. The detailed attributes of the Word2Vec embeddings.

Word Embedding	Language	Property	Dimensionality
GoogleNews-vectors	English	Universal language	300
PTT2015-2018	Chinese	Verbal language	300

4.1. The Performance of User Preference Learning

First, we compared the performance of three classification models: TextCNN, ImageCNN, and early fusion of the two. The training data in Dataset_1 were used to build the models, and the test data in Dateset_1 were used to evaluate the models. In the dataset, there were only user-defined hashtags in addition to texts and images in user posts. It is difficult to know user preferences without explicit user feedback. To obtain the ground truth for user preferences, the hashtag that was used to retrieve each post was set as the class label of that post. We assumed that users implicitly express their preferences through the use of hashtags in the posts and comments that they are interested in.

In our ImageCNN model, the training procedure generally followed Simonyan et al. [45]. In this model, we included three convolutional layers with the kernel size set to 5 and strides = 1. The training was regularized by weight decay (the L2 penalty multiplier was set to 5×10^{-4}) and dropout regularization for the first two fully connected layers with a dropout rate of 0.2. The learning rate was initially set to 0.1. Rectified Linear Unit (ReLU) and Root Mean Square Error (RMSE) were used as the activation function and loss function. The final layer is the softmax layer. The performance of the ImageCNN model is shown in Figure 7.

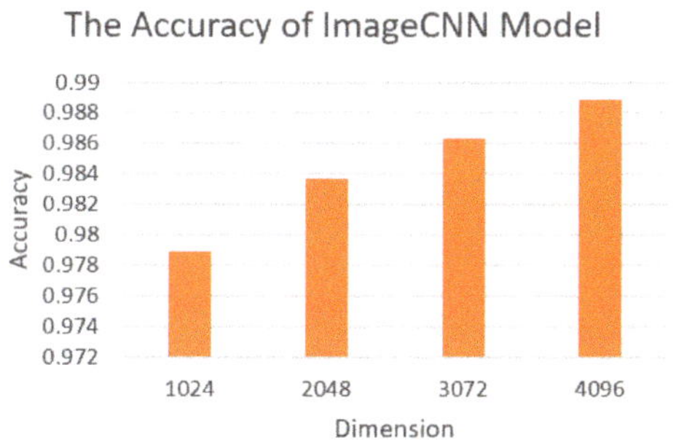

Figure 7. The accuracy of the ImageCNN model.

As shown in Figure 7, when the dimension was 4096, we could obtain the best performance with an accuracy of 98.89%. Since the pretrained model of VGG16 had dimensionality in multiples of 1024, models larger than 4096 could not be run due to the limitation of our computer hardware. In the subsequent experiments, we used the dimension of 4096 for the ImageCNN model.

In our TextCNN model, the training procedure generally followed Kim [41]. We used Word2vec for embedding text contents as input. Three filters of window sizes 3, 4, and 5 were tested, the stride was 1, the dropout rate was set to 0.5, the L2 constraint was 3, and the mini-batch size was set to 50. The learning rate was set to 0.1. We applied a max pooling operation. Training was done through stochastic gradient descent over shuffled mini-batches with the Adadelta update rule [41]. RMSE and LeakyReLu were used as the loss and activation function, respectively. We chose an adaptive learning rate method for neural networks proposed by Geoff Hinton called RMSPROP as the optimizer. The performance of the TextCNN model is shown in Figure 8.

As shown in Figure 8, when the dimension was 2100, we could obtain the best performing TextCNN model with an accuracy of 96.8%. In the subsequent experiments, we used the dimension of 2100 for the TextCNN model. Then, when texts and images were input at the same time, we applied early fusion as shown in Figure 3. The best performance with an accuracy of 80% could be observed when the dimension was 4096. According to our observations, the lower accuracy might have been due to the number of classes in multiclass classification.

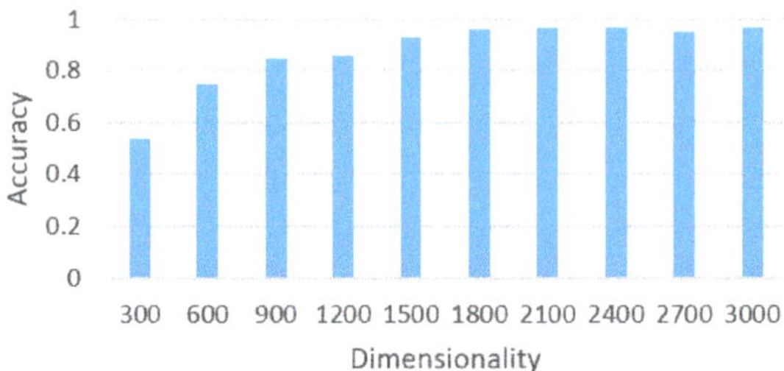

Figure 8. The accuracy of the TextCNN Model.

4.2. The Performance of Similar User Recommendation

To evaluate the performance of user recommendation and to deal with posts in different languages, we further collected a larger dataset (denoted as Dataset_2) where 12 tags were randomly selected from the list of top 100 hashtags on Instagram in 2018 as queries. There were 1143 users and 49,353 posts in Dataset_2, where the major language was Chinese.

In order to compare the performance of different models in similar user recommendation, we established the ground truth for similar users as follows: in the absence of real user preferences, we assumed that the hashtags in each user's posts represents the user's implicit preference and attributes. Thus, the set of hashtags $Tags(u_i)$ for each user u_i was collected from all the posts as the real preference. The similarity between users was calculated by Jaccard similarity as follows:

$$\text{sim}(u_i, u_j) = \text{Jaccard}(\text{Tags}(u_i),\ \text{Tags}(u_j)). \tag{6}$$

Then, the ground truth of top k similar users for a given user u_i was defined as follows:

$$SL_k(u_i) = \{u_j | argmax\_k_j\ sim(u_i, u_j)\}, \tag{7}$$

where $j \neq i$. That is, top k users with the highest similarity to the given user were regarded as the ground truth.

With the recommendation ground truth, we then defined the evaluation metric of the top-k accuracy of a user u_i by calculating the Jaccard similarity between the recommendation list $RL_k(u_i)$ generated by the proposed method and the ground truth $SL_k(u_i)$, as follows:

$$\text{Accuracy}_k(u_i) = \frac{RL_k(u_i) \cap SL_k(u_i)}{RL_k(u_i) \cup SL_k(u_i)}. \tag{8}$$

Then, the overall performance of the model was calculated by the average of the top-k accuracies for all users. To compare the performance of different models, we evaluated the average top-k accuracies for each model when recommending top k similar users, for different values of k.

4.2.1. The Baseline Models

In this paper, we took a different approach to recommender systems than conventional ones. Conventional recommender systems have the premise that the ratings of different users on various items are available. Then, standard recommendation methods such as user-based and item-based collaborative filtering approaches can be used to learn similar users or similar items on the basis of their past rating behaviors, so that ratings for items that were never given before can be deduced. However, in a real-world scenario in social networks where only post content and hashtags are available, it is not possible to obtain the "items" since posts are only posted by one user and replied or shared by a few. It is not possible for them to be "rated" by many different users. In this case, we propose a more realistic approach using hashtags as user preferences for recommending similar users in social networks. When we only have post content and hashtags, the baseline for such a recommendation system is simply content-based recommendation from posts since hashtags have been used as the ground truth.

In our experiments, we considered two baseline models: baseline_text and baseline_pic. The baseline_text model is based on the representation of user preferences only by the text contents of users' posts. Specifically, for each user, we simply use the distributed representation of the texts posted by that user as their characteristics. First, we converted the words into vectors using the same Word2Vec embedding model with the same dictionaries in Table 1. Next, vectors for all words in a document were averaged as the representation of the document. We further averaged the vectors of all documents posted by the user as the user feature. Lastly, we calculated the cosine similarity of the current user to all other users to generate a list of recommendations. On the other hand, the baseline_pic model is based on the representation of user preferences only by the image contents of user posts using the ImageCNN model. The only difference is that we took the featuremaps and categories from all images posted by the user as the user feature.

4.2.2. Effects of the CNN Models

In order to verify the performance of CNN models for extracting text and image features, we compared the average top-k accuracy for the TextCNN model and ImageCNN model with the baseline_text model. First, when using word embedding from Word2Vec for texts, the performance of the baseline_text model on Dataset_1 achieved the average top-k accuracy of 0.128. For the TextCNN model, the average top-k accuracy was 0.326 and 0.322 when the number of dimensions was 2100 and 4096, respectively. For the ImageCNN model, the average top-k accuracy was 0.330 and 0.327 when the number of dimensions was 2100 and 4096, respectively. When we combined the Word2Vec embedded text features and ImageCNN extracted image features by a simple concatenation, the best performance of the baseline_pic model could be obtained with the average top-k accuracy of 0.331. This showed a slightly better performance when using multimodal features than that of individual features. Both TextCNN and ImageCNN models achieved better performance than the baseline_text model. This validates the effectiveness of using CNN models to extract text and image features for finding similar users.

4.2.3. Effects of Multimodal Feature Fusion

To validate the effectiveness of multimodal feature fusion, we further compared the performance of combining text and image features with CNN models using early fusion and late fusion. In early fusion, we concatenate text and image features as inputs to a single CNN model, while, in late fusion, we separately input text features and image features into TextCNN and ImageCNN models, with the resulting feature map and categories are combined as the user feature for recommendation. The performance comparison is shown in Figure 9.

As shown in Figure 9, the best performance of early fusion could be obtained when the number of dimensions was 2100, with the average top-k accuracy of 0.473. For late fusion, the best performance could be obtained with the average top-k accuracy of 0.491, when we used the categories as features, with the number of dimensions of 4096. We could observe the better performance of both early fusion and late fusion than the baseline_pic model with a simple concatenation. This shows the effectiveness of feature fusion, especially for late fusion. In the case of late fusion, we further address the effect of k on the average top-k accuracy and number of hits in Figure 10.

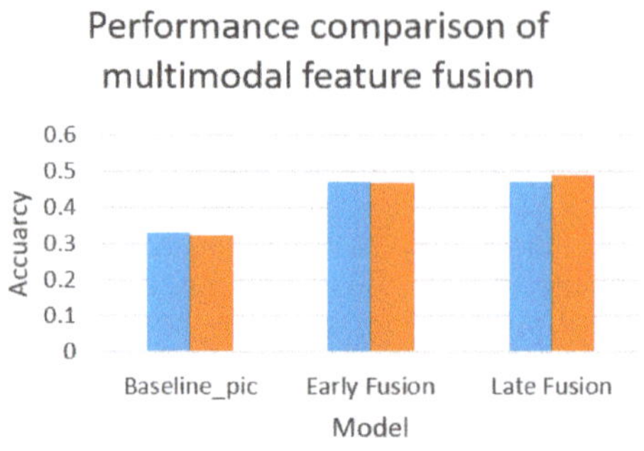

Figure 9. The performance comparison of early and late fusions.

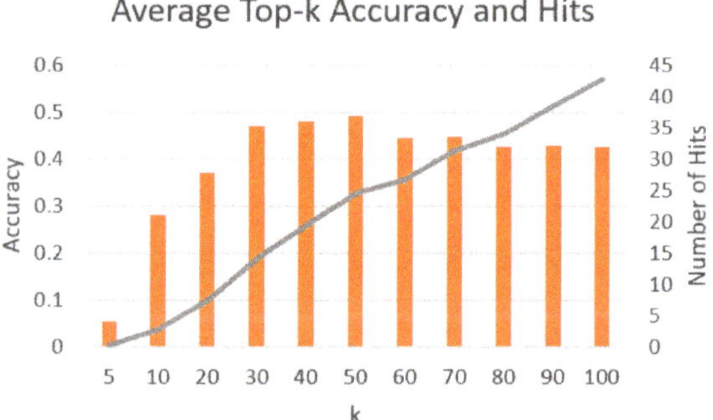

Figure 10. The average top-k accuracy and number of hits for different values of k.

As shown in Figure 10, the average number of hits in our recommendation increased when k increased. Moreover, we could obtain the best average top-k accuracy of 0.491 when k = 50. This was also the best performance in all of our experiments. In the remaining experiments, we used the value of k as 50.

4.2.4. Effects of Autoencoder for Dimension Reduction

To verify the effectiveness of using an autoencoder in finding similar users, we took the features produced by either early fusion or late fusion as inputs to a single-layer neural network for dimension reduction. The performance comparison of autoencoders in different dimensions is shown in Figure 11.

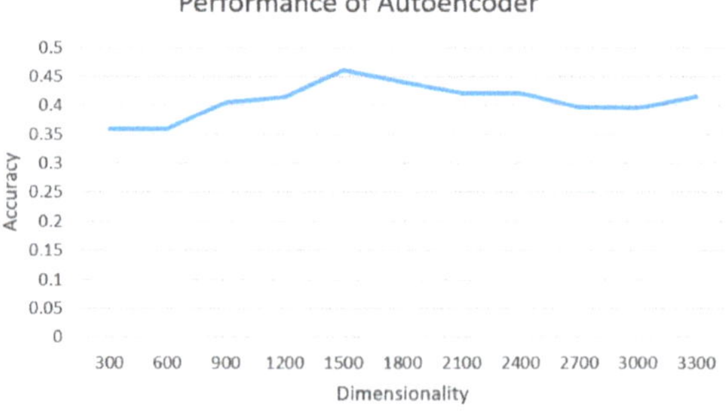

Figure 11. The performance of autoencoders with different dimensionality.

From the experimental results in Figure 11, the best performance could be obtained with the average top-k accuracy of 0.462, when the feature dimension was reduced to 1500 if we used the feature map as the features. This was not as good as the best performance after either early or late fusion. However, the number of dimensions could be greatly reduced from 4096 to 1500. This can help reduce the storage needed for the redundant features. It would be useful to keep the more important features for more efficient computations in practice.

4.2.5. Effects of User Clustering

To verify if user clustering is effective in finding similar users, we further took the feature map and categories from either early fusion or late fusion. Since user clustering can be done offline, only the similarity calculations are needed for those users in the same cluster as the given user. From our experimental results, the best number of clusters was 3 with the average top-k accuracy of 0.367. Since the effectiveness was not as good as a direct recommendation, we further verify the efficiency of recommendation using user clustering in Table 2.

Table 2. The comparison of late fusion and user clustering in terms of execution time. N/A, not applicable.

Model	Number of Recommended Users	Time for Feature Extraction (ms)	Time for Recommendation (ms)	Time for Clustering (ms)	Total Time (ms)
Late fusion	238	2,943,698	507,654	(N/A)	3,451,352
Clustering	238	2,943,698	177,318	1,342,126	4,463,124

As shown in Table 2, although the total amount of time needed in the user clustering model was longer than late fusion due to the time for clustering, the actual amount of time for recommendation was less than that for late fusion. In practical application scenarios, user clustering only needs to be calculated offline once when there are certain amounts of new content or new users. This makes it much faster in the recommendation stage.

To investigate the effects of our proposed approach on a larger scale, we further conducted experiments on Dataset_2. The performance comparison of different models on both datasets is summarized below.

As shown in Table 3, multimodal feature fusion is useful since we can combine clues from both text and image contents. The best performance could be obtained for the late fusion model, with the average top-k accuracy of 0.491 and 0.281 for Dataset_1 and Dataset_2, respectively. Although the accuracy could not be further improved by autoencoders, we could obtain comparable performance with the benefits of dimensionality reduction since a lower dimension means less storage needed, as well as less time required for recommendation systems.

Table 3. The performance comparison of all experimental results.

Model	Dimension	Performance of Dataset_1	Performance of Dataset_2
Baseline_text	300	0.128	0.098
Baseline_pic	2100	0.331	0.130
TextCNN	2100	0.326	0.172
ImageCNN	4096	0.327	0.173
Early fusion	2100	0.473	0.223
Late fusion	4096	0.491	0.281
Autoencoder	1500	0.462	0.245
User clustering	4096	0.367	0.207

5. Discussions

From these experimental results, there are several observations. First, the performance of the CNN-based models (either text or image) is better than the baseline model (text). The improvement of TextCNN or ImageCNN over the baseline is 19% in terms of accuracy. This shows that neural network-based models can improve the performance of content-based baseline models.

However, in this paper, we focused on the task of user recommendation. Existing recommender systems focus on item recommendation on the basis of user ratings on different items. User recommendation is more challenging since we have many users, where each one has different attributes and preferences that cannot be exactly the same as others. When we categorize user preferences, multiclass classification is more difficult than binary classification. To the best of our knowledge, there is no publicly available dataset for user recommendation. Since there is no ground truth in real-world data, we assume that hashtags posted by users represent their implicit user preferences. For each user, the users with more similar hashtags are assumed to be the ground truth of their similar users. Then, we evaluate the performance by the average top-k accuracy of each system-generated user list, as shown in Equation (8). As shown in Section 4, all four proposed model variations outperformed the baseline model. Among the four different variations, late fusion achieved the best performance of 0.491 in terms of average top-k accuracy for similar user recommendation. This shows the effectiveness of the proposed model. Although autoencoders cannot further improve the accuracy, comparable performance can be obtained with the benefits of dimensionality reduction. This gives better efficiency for real-world recommendation systems.

To further assess the performance on a larger scale, all models were executed on Dataset_2. We applied the same parameters as obtained in the best performing models in previous experiments for Dataset_2. From the experimental results, the best performance could be obtained with the average top-k accuracy of 0.281 for late fusion when we used the categories as the user feature with a dimension of 4096. It is worth noting that the proposed approach is far better than the baseline. The improvement across all models is about 8–18%. This further validates the usefulness of our proposed method in practice. The attributes in Dataset_2 are completely different from those in Dataset_1 since they were written in different languages. This shows the effectiveness of our proposed approach in different languages.

The reasons why the performance of Dataset 2 was inferior to that of Dataset 1 are as follows: first, since more tags were used as our queries to Instagram to collect our Dataset 2, we could obtain more posts by more people which could give us more diverse content. Specifically, we included hashtags in Chinese, which involves cross-lingual issues when we cannot effectively identify different hashtags in different languages with similar meanings. One of the reasons why the top-k accuracy for user recommendation was not as high as accuracy in classification tasks is that user-provided hashtags are very diverse. Since our approach is evaluated on the basis of hashtags, this shows its limitation. However, as shown in the experimental results, our proposed approach can outperform the cases of single-modal feature. We demonstrated the effectiveness of multimodal feature fusion from texts and images for user recommendation. In the future, we plan to resolve the issue of diverse hashtags by consolidating the semantic meanings of hashtags using word embedding models or state-of-the-art deep learning models for linguistic tasks such as transformers or Bidirectional Encoder Representations from Transformers (BERT).

As a potential application of our proposed approach, we can utilize the similar user recommendation algorithm to build an intellectual data crawler in social networks. For example, according to the targeted topics of interest, we can utilize our proposed approach to discover related posts, with multimedia content and related user information. Then, by clustering similar users on the basis of user preferences, it would be useful to further expand our proposed approach across multiple social networks that might be different in their structure. This could help reduce the problem of social network analysis across different social networks.

In real application scenarios, there could be issues such as posted and reposted images, as well as drawings and photos of the same thing, to name a few. In this paper, we did not distinguish between posted and reposted images if they were captured in the same resolution. Since the features of all images were extracted by the same CNN architecture of VGG16, the posted and reposted images would have the same features if they are represented by the same pixels. This classifies them into the same category, from which our recommendation is made. Following the same line of thought, various content such as drawings and photos of the same thing would not have exactly the same features since it would be difficult to mimic the photos when drawing the same thing. However, CNN models were demonstrated [52] to improve the performance of sketch-based image retrieval (SBIR) by extracting deep features in recent years. Since we utilize VGG16 with multiple convolutional and pooling layers, our algorithm is able to extract the important semantic features from drawings and photos of the same thing that give similar classification results.

6. Conclusions

In this paper, we proposed a multimodal feature fusion approach to user preference learning by combining user post contents and related image features for recommending similar users on a popular social platform, Instagram. With the help of convolutional networks, the features of images and texts can be effectively extracted. Using either early or late fusion methods, these features can be effectively integrated as a representation of user preference. According to the experimental results, our proposed method achieved good performance on the real-world datasets. Firstly, the effectiveness of our scheme was better than the conventional method of distributed representation of texts only. Secondly, the late fusion approach combining images and texts obtained the best performance of 0.491 in terms of average top-k accuracy. Thirdly, the proposed method could also be applied in datasets with different language attributes. This shows the effectiveness and potential of our proposed approach to similar user recommendation in social networks. Further investigation is needed to verify the performance of our proposed approach on different social media platforms.

Author Contributions: Conceptualization, J.-H.W. and L.W.; methodology, Y.-T.W.; software, Y.-T.W.; writing—original draft preparation, Y.-T.W.; writing—review and editing, J.-H.W. and L.W. All authors read and agreed to the published version of the manuscript.

Funding: This work was partially funded by the research grants of the Ministry of Science and Technology, Taiwan under the grant number MOST109-2221-E-027-090 and partially funded by the National Taipei University of Technology–University of Science and Technology Beijing Joint Research Program under the grant number of NTUT-USTB-107-07. It was also partially funded by the National Applied Research Laboratories, Taiwan under the grant number of NARL- ISIM-109-002.

Institutional Review Board Statement: Not applicable.

Informed Consent Statement: Not applicable.

Data Availability Statement: The data presented in this study are available on request from the corresponding author. The data are not publicly available due to privacy concerns as stated in Instagram Data Policy.

Conflicts of Interest: The authors declare no conflict of interest.

References

1. Chen, J.; Geyer, W.; Dugan, C.; Muller, M.; Guy, I. "Make New Friends, but Keep the Old"–Recommending People on Social Networking Sites. In Proceedings of the SIGCHI Conference on Human Factors in Computing Systems (CHI'09), Boston, MA, USA, 4–9 April 2009; pp. 201–210.
2. Hannon, J.; Bennett, M.; Smyth, B. Recommending Twitter Users to Follow Using Content and Collaborative Filtering Approaches. In Proceedings of the fourth ACM Conference on Recommender Systems (RecSys'10), Barcelona, Spain, 26–30 September 2010; pp. 199–206.
3. Armentano, M.G.; Godoy, D.; Amandi, A. Topology-based recommendation of users in micro-blogging communities. *J. Comput. Sci. Technol.* **2012**, *27*, 624–634. [CrossRef]
4. Zhao, G.; Lee, M.L.; Hsu, W.; Chen, W.; Hu, H. Community-Based User Recommendation in Uni-Directional Social Networks. In Proceedings of the 22nd ACM international conference on Information & Knowledge Management (CIKM'13), San Francisco, CA, USA, 27 October–1 November 2013; pp. 189–198.

5. Gurini, D.F.; Gasparetti, F.; Micarelli, A.; Sansonetti, G. Temporal people-to-people recommendation on social networks with sentiment-based matrix factorization. *Future Gener. Comput. Syst.* **2018**, *78*, 430–439. [CrossRef]
6. Lee, R.K.-W.; Hoang, T.-A.; Lim, E.-P. On analyzing user topic-specific platform preferences across multiple social media sites. In Proceedings of the 26th International Conference on World Wide Web, Perth, Australia, 3–7 April 2017; pp. 1351–1359.
7. Li, X.; Pham, T.-A.N.; Cong, G.; Yuan, Q.; Li, X.-L. Krishnaswamy S Where you instagram? Associating your instagram photos with points of interest. In Proceedings of the 24th ACM International on Conference on Information and Knowledge Management, Melbourne, Australia, 19–23 October 2015; pp. 1231–1240.
8. Caicedo, J.C.; Kapoor, A.; Kang, S.B. Collaborative personalization of image enhancement. In Proceedings of the CVPR 2011, Providence, RI, USA, 20–25 June 2011; pp. 249–256.
9. Wei, H.; Zhang, F.; Yuan, N.J.; Cao, C.; Fu, H.; Xie, X.; Rui, Y.; Ma, W.-Y. Beyond the words: Predicting user personality from heterogeneous information. In Proceedings of the Tenth ACM International Conference on Web Search and Data Mining, Cambridge, UK, 6–10 February 2017; pp. 305–314.
10. Long, Y.; Lu, Q.; Xiao, Y.; Li, M.; Huang, C.-R. Domain-specific user preference prediction based on multiple user activities. In Proceedings of the 2016 IEEE International Conference on Big Data (Big Data), Washington, DC, USA, 5–8 December 2016; pp. 3913–3921.
11. Volkova, S.; Coppersmith, G.; Van Durme, B. Inferring user political preferences from streaming communications. In Proceedings of the 52nd Annual Meeting of the Association for Computational Linguistics, Baltimore, MD, USA, 22–27 June 2014; Volume 1, pp. 186–196.
12. Yousif, A.; Niu, Z.; Tarus, J.K.; Ahmad, A. A survey on sentiment analysis of scientific citations. *Artif. Intell. Rev.* **2019**, *52*, 1805–1838. [CrossRef]
13. Nilashi, M.; Ahani, A.; Esfahani, M.D.; Yadegaridehkordi, E.; Samad, S.; Ibrahim, O.; Sharef, N.M.; Akbari, E. Preference learning for eco-friendly hotels recommendation: A multi-criteria collaborative filtering approach. *J. Clean. Prod.* **2019**, *215*, 767–783. [CrossRef]
14. Tarus, J.K.; Niu, Z.; Kalui, D. A hybrid recommender system for e-learning based on context awareness and sequential pattern mining. *Soft Comput.* **2018**, *22*, 2449–2461. [CrossRef]
15. Wan, S.; Niu, Z. An e-learning recommendation approach based on the self-organization of learning resource. *Knowl. Based Syst.* **2018**, *160*, 71–87. [CrossRef]
16. Wan, S.; Niu, Z. A Hybrid E-learning Recommendation Approach Based on Learners' Influence Propagation. *IEEE Trans. Knowl. Data Eng.* **2019**, *32*, 827–840. [CrossRef]
17. Zhang, Y.; Lai, G.; Zhang, M.; Zhang, Y.; Liu, Y.; Ma, S. Explicit factor models for explainable recommendation based on phrase-level sentiment analysis. In Proceedings of the 37th international ACM SIGIR Conference on Research & Development in Information Retrieval, Gold Coast, QLD, Australia, 6–11 July 2014; pp. 83–92.
18. Barkan, O. Bayesian neural word embedding. In Proceedings of the Thirty-First AAAI Conference on Artificial Intelligence, San Francisco, CA, USA, 4–9 February 2017; pp. 3135–3143.
19. He, X.; Liao, L.; Zhang, H.; Nie, L.; Hu, X.; Chua, T.-S. Neural collaborative filtering. In Proceedings of the 26th International Conference on World Wide Web, Perth, Australia, 3–7 April 2017; pp. 173–182.
20. Palangi, H.; Deng, L.; Shen, Y.; Gao, J.; He, X.; Chen, J.; Song, X.; Ward, R. Deep sentence embedding using long short-term memory networks: Analysis and application to information retrieval. *IEEE/ACM Trans. Audio Speech Lang. Process.* **2016**, *24*, 694–707. [CrossRef]
21. Tai, K.S.; Socher, R.; Manning, C.D. Improved semantic representations from tree-structured long short-term memory networks. In Proceedings of the Annual Meeting of the Association for Computational Linguistics, Beijing, China, 26–31 July 2015.
22. Yousif, A.; Niu, Z.; Chambua, J.; Khan, Z.Y. Multi-task learning model based on recurrent convolutional neural networks for citation sentiment and purpose classification. *Neurocomputing* **2019**, *335*, 195–205. [CrossRef]
23. Seo, S.; Huang, J.; Yang, H.; Liu, Y. Representation learning of users and items for review rating prediction using attention-based convolutional neural networks. In Proceedings of the SIAM Conference on Data Mining 2017, Houston, Texas, USA, 27–29 April 2017.
24. Chambua, J.; Niu, Z.; Zhu, Y. User preferences prediction approach based on embedded deep summaries. *Expert Syst. Appl.* **2019**, *132*, 87–98. [CrossRef]
25. Lv, J.; Liu, W.; Zhang, M.; Gong, H.; Wu, B.; Ma, H. Multi-feature fusion for predicting social media popularity. In Proceedings of the 25th ACM International Conference on Multimedia, Mountain View, CA, USA, 23–27 October 2017; pp. 1883–1888.
26. Zhang, W.; Wang, W.; Wang, J.; Zha, H. User-guided hierarchical attention network for multi-modal social image popularity prediction. In Proceedings of the 2018 World Wide Web Conference, International World Wide Web Conferences Steering Committee, Lyon, France, 23–27 April 2018; pp. 1277–1286.
27. Aloufi, S.; Zhu, S.; El Saddik, A. On the prediction of flickr image popularity by analyzing heterogeneous social sensory data. *Sensors* **2017**, *17*, 631. [CrossRef] [PubMed]
28. Mazloom, M.; Hendriks, B.; Worring, M. Multimodal context-aware recommender for post popularity prediction in social media. In Proceedings of the Thematic Workshops of ACM Multimedia 2017, Mountain View, CA, USA, 23–27 October 2017; pp. 236–244.
29. Borth, D.; Ji, R.; Chen, T.; Breuel, T.; Chang, S.-F. Large-scale visual sentiment ontology and detectors using adjective noun pairs. In Proceedings of the 21st ACM International Conference on Multimedia 2013, Barcelona, Spain, 21–25 October 2013; pp. 223–232.

30. Thelwall, M.; Buckley, K.; Paltoglou, G.; Cai, D.; Kappas, A. Sentiment strength detection in short informal text. *J. Am. Soc. Inf. Sci. Technol.* **2010**, *61*, 2544–2558. [CrossRef]
31. Donoser, M. Bischof H Diffusion processes for retrieval revisited. In Proceedings of the IEEE Conference on Computer Vision and Pattern Recognition, Portland, OR, USA, 23–28 June 2013; pp. 1320–1327.
32. Dehghani, M.; Zamani, H.; Severyn, A.; Kamps, J.; Croft, W.B. Neural ranking models with weak supervision. In Proceedings of the 40th International ACM SIGIR Conference on Research and Development in Information Retrieval, Tokyo, Japan, 7–11 August 2017; pp. 65–74.
33. Gallo, I.; Calefati, A.; Nawaz, S. Multimodal classification fusion in real-world scenarios. In Proceedings of the 2017 14th IAPR International Conference on Document Analysis and Recognition (ICDAR), Kyoto, Japan, 9–15 November 2017; pp. 36–41.
34. Kiela, D.; Grave, E.; Joulin, A.; Mikolov, T. Efficient large-scale multi-modal classification. In Proceedings of the Thirty-Second AAAI Conference on Artificial Intelligence, New Orleans, LA, USA, 2–7 February 2018; pp. 5198–5204.
35. Baltrušaitis, T.; Ahuja, C.; Morency, L.-P. Multimodal machine learning: A survey and taxonomy. *IEEE Trans. Pattern Anal. Mach. Intell.* **2018**, *41*, 423–443. [CrossRef] [PubMed]
36. Lan, Z.-Z.; Bao, L.; Yu, S.-I.; Liu, W.; Hauptmann, A.G. Multimedia classification and event detection using double fusion. *Multimedia Tools Appl.* **2013**, *71*, 333–347. [CrossRef]
37. Shutova, E.; Kiela, D.; Maillard, J. Black holes and white rabbits: Metaphor identification with visual features. In Proceedings of the 2016 Conference of the North American Chapter of the Association for Computational Linguistics: Human Language Technologies, San Diego, CA, USA, 12–17 June 2016; pp. 160–170.
38. Morvant, E.; Habrard, A.; Ayache, S. Majority vote of diverse classifiers for late fusion. In *Joint IAPR International Workshops on Statistical Techniques in Pattern Recognition (SPR) and Structural and Syntactic Pattern Recognition (SSPR)*; Springer: Berlin/Heidelberg, Germany, 2014; pp. 153–162.
39. Lv, P.; Meng, X.; Zhang, Y. FeRe: Exploiting influence of multi-dimensional features resided in news domain for recommendation. *Inf. Process. Manag.* **2017**, *53*, 1215–1241. [CrossRef]
40. Khatibi, A.; Belém, F.M.; Da Silva, A.P.C.; Almeida, J.; Gonçalves, M. Fine-grained tourism prediction: Impact of social and environmental features. *Inf. Process. Manag.* **2020**, *57*, 102057. [CrossRef]
41. Kim, Y. Convolutional neural networks for sentence classification. *arXiv* **2014**, arXiv:14085882.
42. Mikolov, T.; Sutskever, I.; Chen, K.; Corrado, G.S.; Dean, J. Distributed representations of words and phrases and their compositionality. *Adv. Neural Inf. Process. Syst.* **2013**, arXiv:1310.4546.
43. Mikolov, T.; Chen, K.; Corrado, G.; Dean, J. Efficient estimation of word representations in vector space. *arXiv* **2013**, arXiv:13013781.
44. Collobert, R.; Weston, J.; Bottou, L.; Karlen, M.; Kavukcuoglu, K.; Kuksa, P. Natural language processing (almost) from scratch. *J. Mach. Learn. Res.* **2011**, *12*, 2493–2537.
45. Simonyan, K.; Zisserman, A. Very deep convolutional networks for large-scale image recognition. *arXiv* **2014**, arXiv:14091556.
46. Ciresan, D.C.; Meier, U.; Masci, J.; Gambardella, L.M.; Schmidhuber, J. Flexible, high performance convolutional neural networks for image classification. In Proceedings of the Twenty-Second International Joint Conference on Artificial Intelligence, Barcelona, Catalonia, Spain, 16–22 July 2011; pp. 1237–1242.
47. Krizhevsky, A.; Sutskever, I.; Hinton, G.E. Imagenet classification with deep convolutional neural networks. *Adv. Neural Inf. Process. Syst.* **2012**, 1097–1105.
48. Qassim, H.; Verma, A.; Feinzimer, D. Compressed residual-VGG16 CNN model for big data places image recognition. In Proceedings of the 2018 IEEE 8th Annual Computing and Communication Workshop and Conference (CCWC), Las Vegas, NV, USA, 8–10 January 2018; pp. 169–175.
49. Atrey, P.K.; Hossain, M.A.; El Saddik, A.; Kankanhalli, M.S. Multimodal fusion for multimedia analysis: A survey. *Multimedia Syst.* **2010**, *16*, 345–379. [CrossRef]
50. Caliński, T.; Harabasz, J. A dendrite method for cluster analysis. *Commun. Stat. Theory Methods* **1974**, *3*, 1–27. [CrossRef]
51. Hamerly, G.; Elkan, C. Alternatives to the k-means algorithm that find better clusterings. In Proceedings of the Eleventh International Conference on Information and Knowledge Management, McLean, VA, USA, 4–9 November 2002; pp. 600–607.
52. Yu, D.; Liu, Y.; Pang, Y.; Li, Z.; Li, H. A multi-layer deep fusion convolutional neural network for sketch based image retrieval. *Neurocomputing* **2018**, *296*, 23–32. [CrossRef]

Article

Improvement in the Convolutional Neural Network for Computed Tomography Images

Keisuke Manabe [1], Yusuke Asami [1], Tomonari Yamada [1] and Hiroyuki Sugimori [2,*]

[1] Graduate School of Health Sciences, Hokkaido University, Sapporo 060-0812, Japan; ksk0843@eis.hokudai.ac.jp (K.M.); yusuke12@eis.hokudai.ac.jp (Y.A.); tomonarihandball@eis.hokudai.ac.jp (T.Y.)

[2] Faculty of Health Sciences, Hokkaido University, Sapporo 060-0812, Japan

* Correspondence: sugimori@hs.hokudai.ac.jp; Tel.: +81-11-706-3410

Abstract: Background and purpose. This study evaluated a modified specialized convolutional neural network (CNN) to improve the accuracy of medical images. Materials and Methods. We defined computed tomography (CT) images as belonging to one of the following 10 classes: head, neck, chest, abdomen, and pelvis with and without contrast media, with 10,000 images per class. We modified the CNN based on the AlexNet with an input size of 512 × 512. We resized the filter sizes of the convolution layer and max pooling. Using these modified CNNs, various models were created and evaluated. The improved CNN was evaluated to classify the presence or absence of the pancreas in the CT images. We compared the overall accuracy, which was calculated from images not used for training, to that of the ResNet. Results. The overall accuracies of the most improved CNN and ResNet in the 10 classes were 94.8% and 89.3%, respectively. The filter sizes of the improved CNN for the convolution layer were (13, 13), (7, 7), (5, 5), (5, 5), and (5, 5) in order from the first layer, and that of max-pooling was (7, 7). The calculation times of the most improved CNN and ResNet were 56 and 120 min, respectively. Regarding the classification of the pancreas, the overall accuracies of the most improved CNN and ResNet were 75.75% and 58.25%, respectively. The calculation times of the most improved CNN and ResNet were 36 and 55 min, respectively. Conclusion. By optimizing the filter size of the convolution layer and max-pooling of 512 × 512 images, we quickly obtained a highly accurate medical image classification model. This improved CNN can be useful for classifying lesions and anatomies for related diagnostic aid applications.

Keywords: deep learning; convolutional neural network; image classification

Citation: Manabe, K.; Asami, Y.; Yamada, T.; Sugimori, H. Improvement in the Convolutional Neural Network for Computed Tomography Images. *Appl. Sci.* **2021**, *11*, 1505. https://doi.org/10.3390/app11041505

Academic Editor: Chuan-Ming Liu
Received: 12 January 2021
Accepted: 5 February 2021
Published: 7 February 2021

Publisher's Note: MDPI stays neutral with regard to jurisdictional claims in published maps and institutional affiliations.

Copyright: © 2021 by the authors. Licensee MDPI, Basel, Switzerland. This article is an open access article distributed under the terms and conditions of the Creative Commons Attribution (CC BY) license (https://creativecommons.org/licenses/by/4.0/).

1. Introduction

Image classification is a typical technology of image analysis that uses artificial intelligence. In computed tomography (CT) images, it has been used to classify pulmonary nodules [1,2], slice positions [3,4], and calcaneal fractures [5,6]. Both AlexNet [7] and ResNet [8] are examples of image classification models. Because AlexNet has fewer layers than ResNet, its accuracy is low but the calculation time is short. Learning images with a large pixel size results in graphics processing unit (GPU) memory shortage, as the calculation cost is large. Therefore, images resized to 224 × 224, which is the default pixel size of most image classification models, are often used for training [9,10]. In the study by Santin et al. [9], augmentation of data and transfer learning were used to improve the accuracy and robustness of the model, but the pixel size of the input images was 224 × 224 × 3; thus, the pixel size was not examined. However, in medical images, the general pixel size is 512 × 512. Reducing the image size may reduce the number of features as a result of the compression of image information. Because several tens of thousands of images are required for medical image classification, learning takes a lot of time [11]. The existing image classification model, AlexNet, has low accuracy because it is an early model, but the calculation time is short because it has few layers. If AlexNet can be specialized to

512 × 512, two problems may be solved: the reduction of features due to resizing and the long calculation time, which is a weakness of models with a large number of parameters. Being able to train with the original size images is useful because, in actual diagnosis, micro lesions need to be detected. In this study, the parameters of AlexNet were customized for medical images, and the accuracy and calculation time of the convolutional neural network (CNN) were evaluated. The generalization capability of the improved CNN was evaluated by classifying the presence or absence of the pancreas, which is considered difficult [12].

2. Materials and Methods

2.1. Subjects and Datasets

In this study, we targeted 118,000 512 × 512 axial CT images. These images were approved by the Hokkaido University Hospital Ethics Committee. In the 10 classes for adjusting AlexNet parameters, 100,000 images were used for training, and 10,000 images were used for accuracy verification. In the classification of the presence or absence of the pancreas to evaluate the generalization capability of the improved CNN, 6000 images were used for training, and 2000 images were used for accuracy verification.

2.2. Ten Classes for Adjusting AlexNet Parameters

We defined training and accuracy verification images as the following 10 classes: head, neck, chest, abdomen, and pelvis with and without contrast media. Training images comprised 10,000 images per class, and accuracy verification images included 1000 per class. The original AlexNet was trained with 224 × 224 and 512 × 512 images, and the overall accuracy and calculation time of these two models were compared. The respective models were named original (input image size: 224 × 224) and original (input image size: 512 × 512). The AlexNet parameters adjusted for 512 × 512 images were the filter sizes of the convolution layer and max pooling. Because the filter sizes of the convolution layer had many change patterns, we divided them into five groups—A, B, C, D, and E—and named each model as A1 and A2 and so on. Figure 1 presents the change patterns of groups A to E. Figure 2 shows the original and changed values of the filter size of the convolution layer. On the other hand, the original value of the filter size of max-pooling was (3, 3), which we changed to an odd number of 5 to 15. We calculated the overall accuracies of these models using accuracy verification images. Various models were created with all combinations of parameters exceeding the overall accuracy of the original (input image size: 512 × 512), and the overall accuracy was calculated. The names of these models were "group name of convolution layer"–"filter size of max pooling," such as A1-5. Next, the original ResNet was trained with 224 × 224 images, and the overall accuracy and calculation time were calculated. We compared the overall accuracy, confusion matrix, and calculation time of the model with the highest overall accuracy among the created models to those of the ResNet.

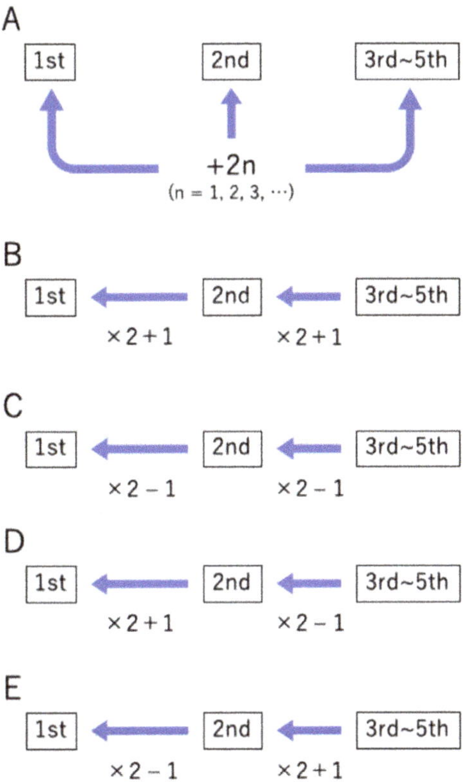

Figure 1. Change patterns of the filter size of the convolution layer. [(**A**–**E**): the change patterns of group name (groups **A** to **E**)].

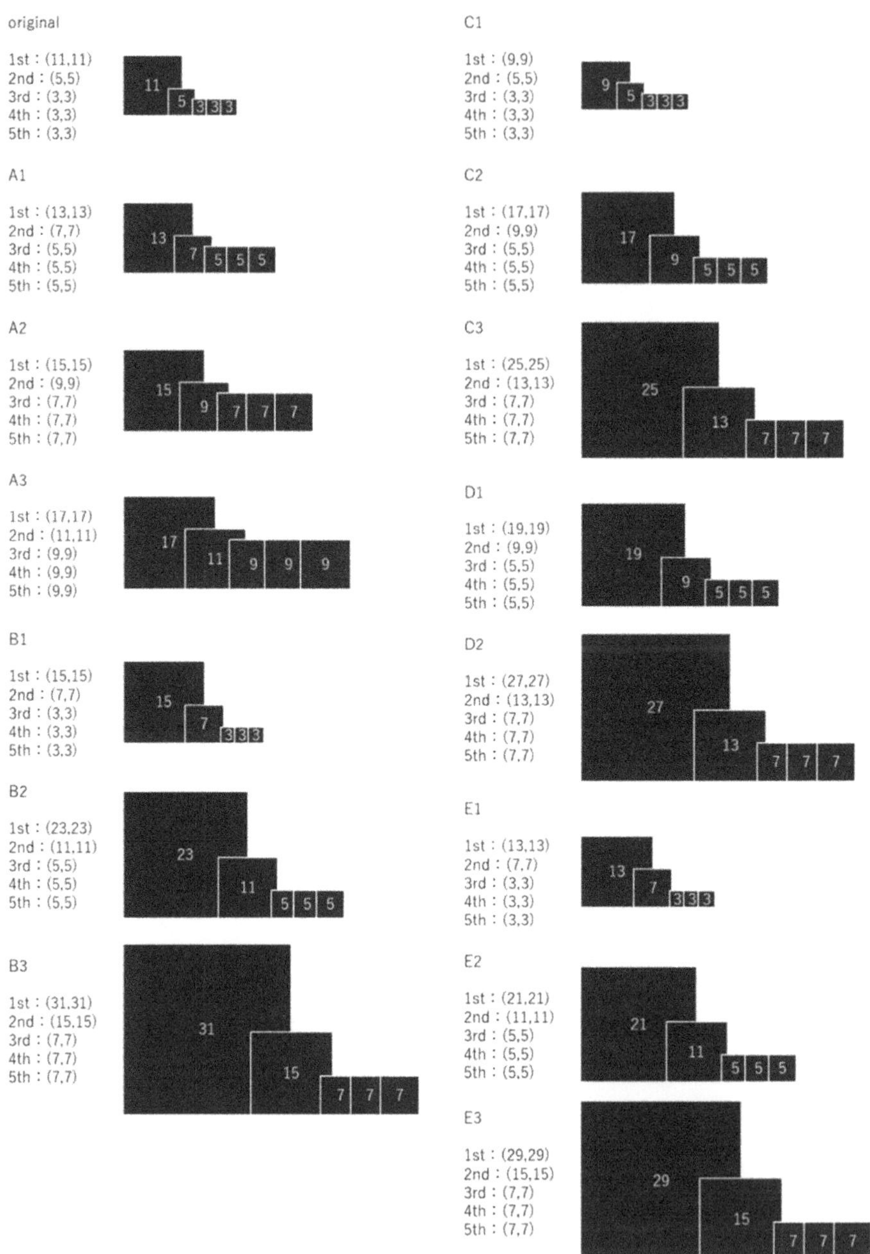

Figure 2. Original and changed values of the filter size of the convolution layer. [(**A**–**E**): the change patterns of group name (groups **A** to **E**)].

2.3. Classification of the Presence or Absence of the Pancreas to Evaluate the Generalization Capability of the Improved CNN

We defined training and accuracy verification images as the following four classes: the presence or absence of the pancreas with and without contrast media. Training images were 1500 images per class, and accuracy verification images were 500 per class. ResNet and the model with the highest overall accuracy among the created models were trained, and we compared the overall accuracy, confusion matrix, and calculation time. Figure 3 shows the entire learning and evaluation process. For the training, we used a PC with NVIDIA GeForce GTX TITAN X 12GB (NVIDIA Corporation, Santa Clara, CA, USA).

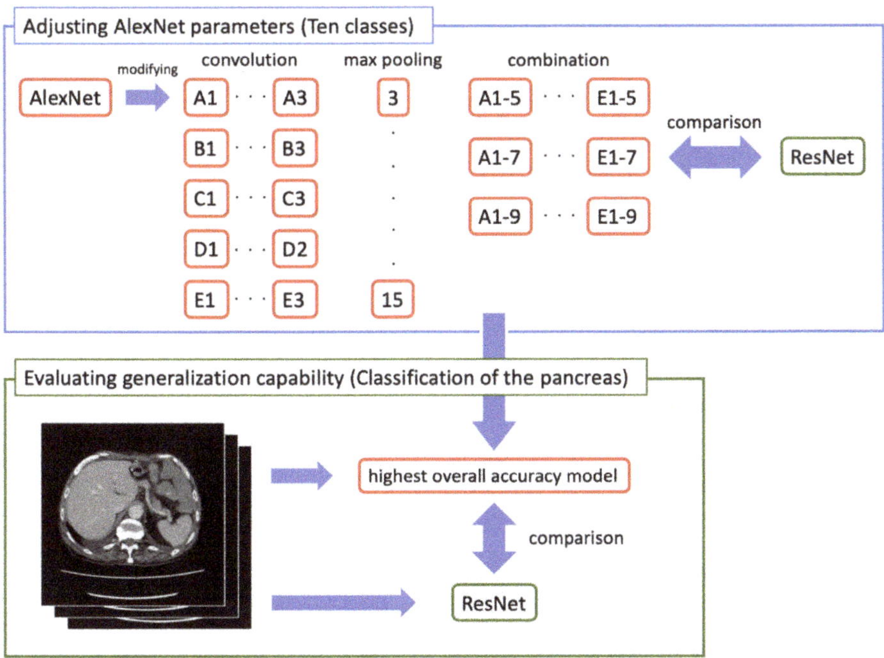

Figure 3. The entire learning and evaluation process.

3. Results

3.1. Ten Classes for Adjusting AlexNet Parameters

Table 1 shows the overall accuracies and calculation times of the original AlexNet trained with 224 × 224 and 512 × 512 images. In the comparison between original (input image size: 224 × 224) and original (input image size: 512 × 512), original (input image size: 512 × 512) had higher overall accuracy and a shorter calculation time. Figure 4 presents the overall accuracies and calculation times of the models in which the filter sizes of the convolution layer were changed, and Figure 5 shows the overall accuracies and calculation times of the models in which the filter sizes of max-pooling were changed. The models exceeding the overall accuracy of original (input image size: 512 × 512) were A1~3, B1~3, C1~2, D1~2, and E1, and the models with filter sizes of max-pooling were 5, 7, and 9. Figures 6–8 show the overall accuracies and calculation times of the models with all combinations of these parameters. Among these models, the highest overall accuracy was in model A1-7. The filter sizes of the convolution layer for A1-7 were (13, 13), (7, 7), (5, 5), (5, 5), and (5, 5) in order from the first layer, and that of max-pooling was (7, 7). The overall accuracy of A1-7 was 94.40%, and the calculation time was 56 min. Figure 9 displays the confusion matrix. In contrast, the overall accuracy of ResNet was 88.80%, and calculation time was 120 min. The confusion matrix is shown in Figure 10. In the comparison of A1-7 and ResNet, A1-7 was superior in both overall accuracy and calculation time.

Table 1. Overall accuracies and calculation times of the original AlexNet.

	Overall Accuracy [%]	Calculation Time [min]
Original (input image size: 224 × 224)	82.5	41
Original (input image size: 512 × 512)	86.7	29

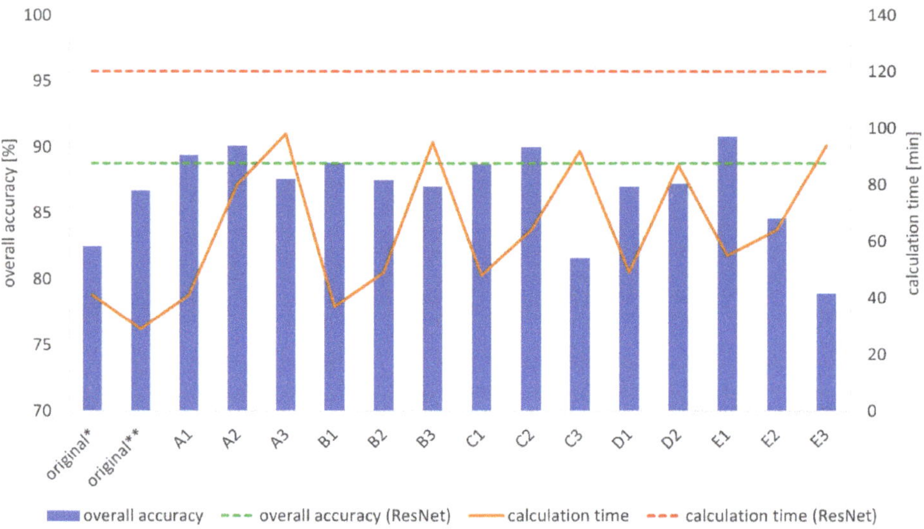

Figure 4. Overall accuracies and calculation times of the models in which the filter sizes of the convolution layer were changed (original *: original (input image size: 224 × 224), original **: original (input image size: 512 × 512)).

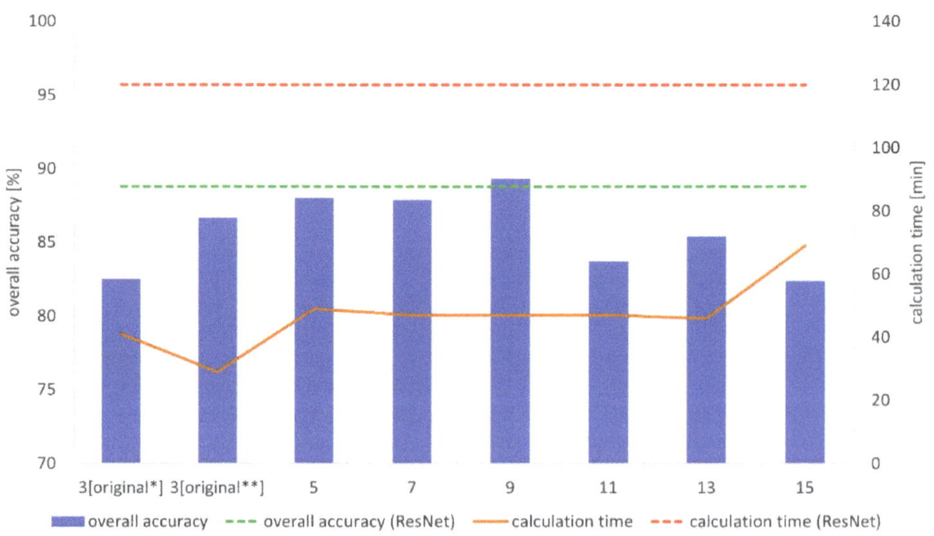

Figure 5. Overall accuracies and calculation times of the models with the filter sizes of max-pooling (original *: original (input image size: 224 × 224), original **: original (input image size: 512 × 512)).

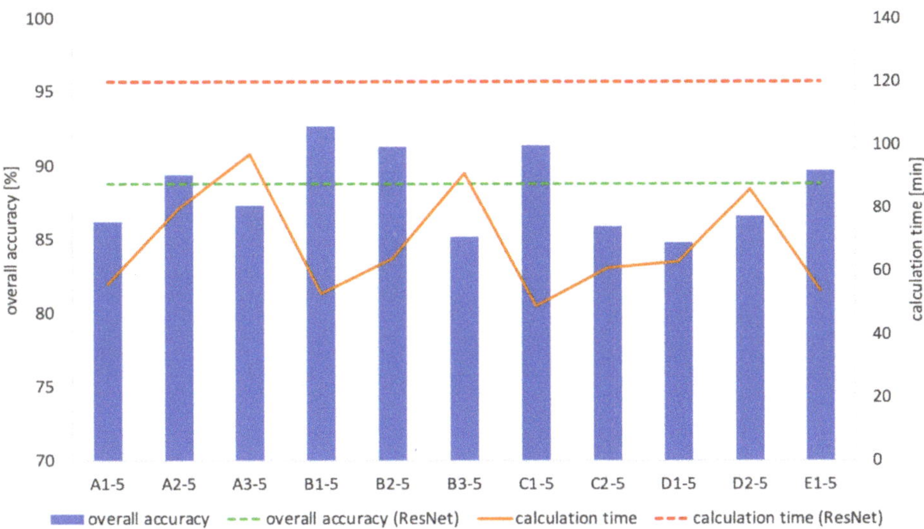

Figure 6. Overall accuracies and calculation times of the models with combination (the filter size of max-pooling: (5, 5)).

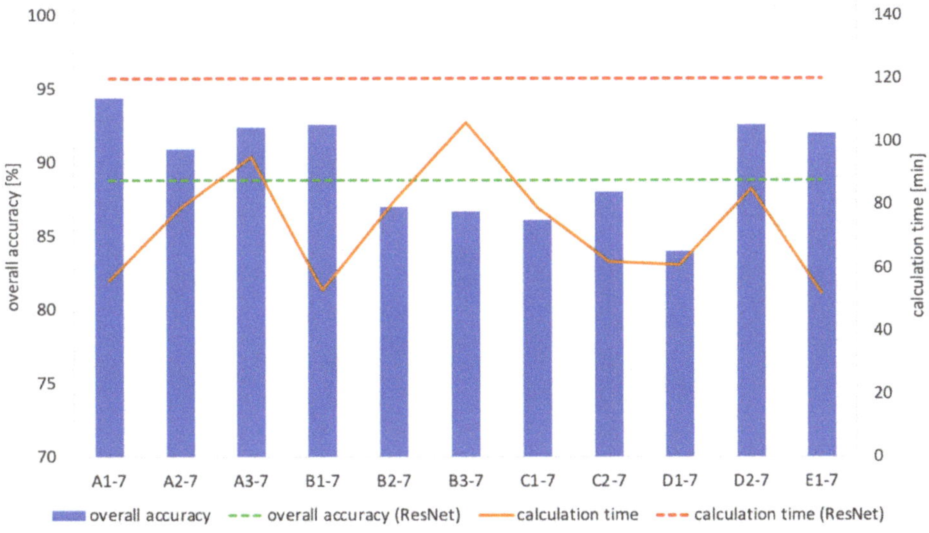

Figure 7. Overall accuracies and calculation times of the models with combination (the filter size of max-pooling: (7, 7)).

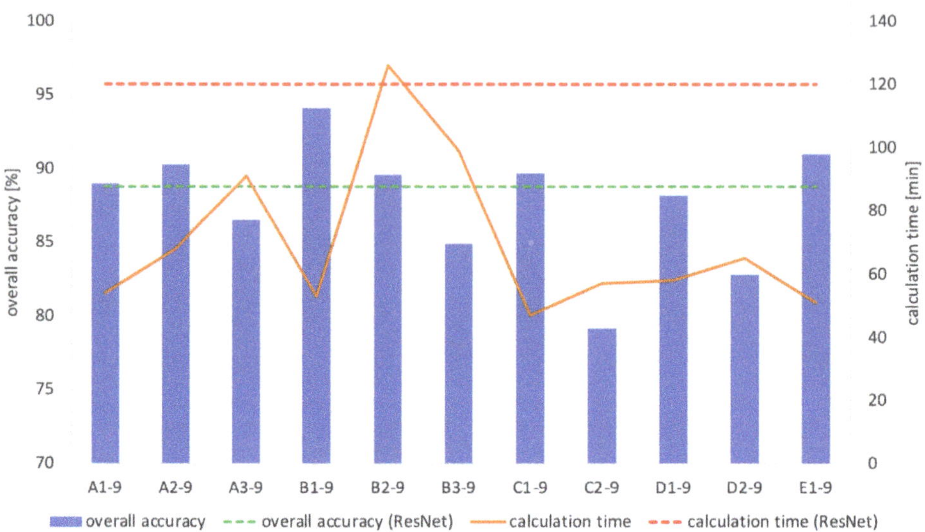

Figure 8. Overall accuracies and calculation times of the models with combination (the filter size of max-pooling: (9, 9)).

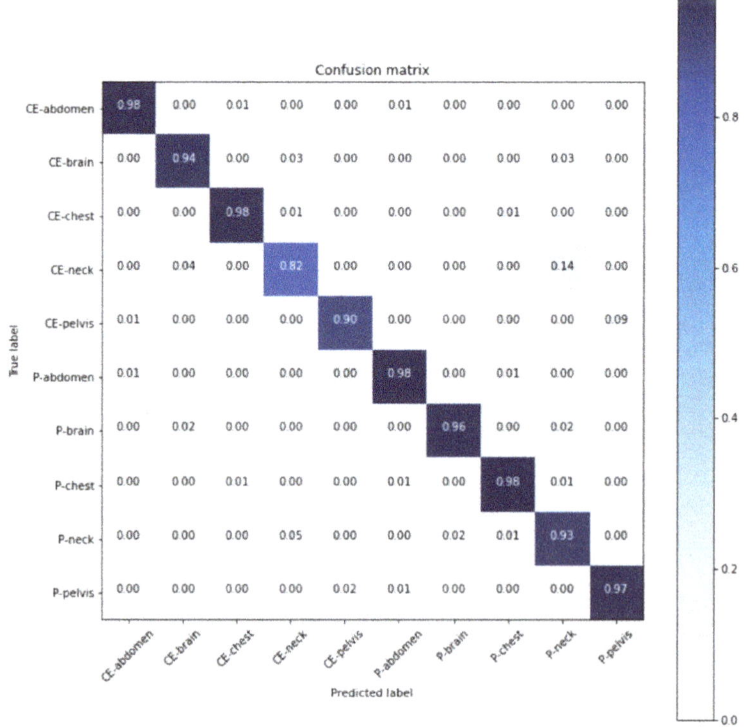

True label	CE-abdomen	CE-brain	CE-chest	CE-neck	CE-pelvis	P-abdomen	P-brain	P-chest	P-neck	P-pelvis
CE-abdomen	0.98	0.00	0.01	0.00	0.00	0.01	0.00	0.00	0.00	0.00
CE-brain	0.00	0.94	0.00	0.03	0.00	0.00	0.00	0.00	0.03	0.00
CE-chest	0.00	0.00	0.98	0.01	0.00	0.00	0.00	0.01	0.00	0.00
CE-neck	0.00	0.04	0.00	0.82	0.00	0.00	0.00	0.00	0.14	0.00
CE-pelvis	0.01	0.00	0.00	0.00	0.90	0.00	0.00	0.00	0.00	0.09
P-abdomen	0.01	0.00	0.00	0.00	0.00	0.98	0.00	0.01	0.00	0.00
P-brain	0.00	0.02	0.00	0.00	0.00	0.00	0.96	0.00	0.02	0.00
P-chest	0.00	0.00	0.01	0.00	0.00	0.01	0.00	0.98	0.01	0.00
P-neck	0.00	0.00	0.00	0.05	0.00	0.00	0.02	0.01	0.93	0.00
P-pelvis	0.00	0.00	0.00	0.00	0.02	0.01	0.00	0.00	0.00	0.97

Figure 9. Confusion matrix of A1-7. (CE: contrast-enhanced, P: plain).

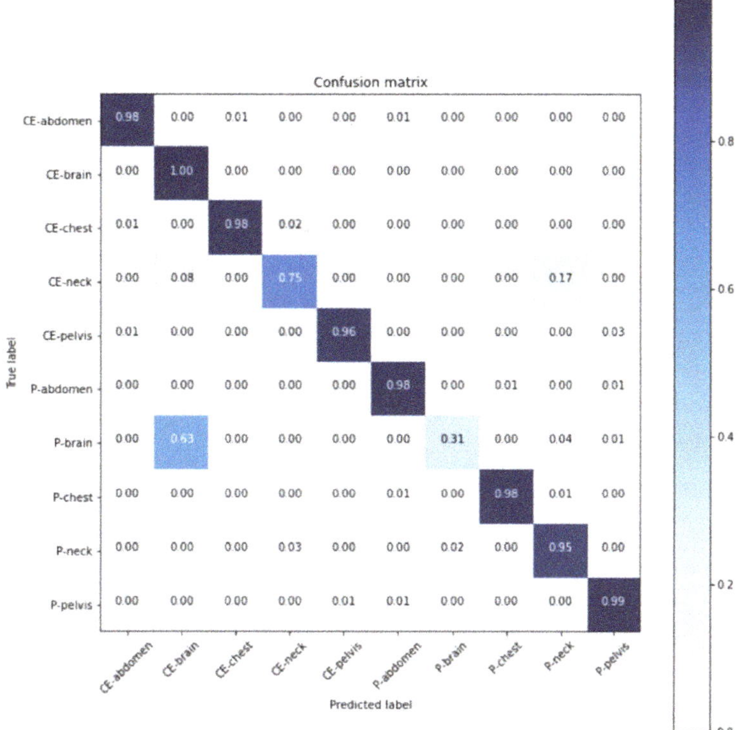

Figure 10. Confusion matrix of ResNet (CE: contrast-enhanced, P: plain).

3.2. Classification of the Presence or Absence of the Pancreas for the Evaluation of the Generalization Capability of the Improved CNN

The overall accuracy of A1-7 was 75.75%, and the calculation time was 36 min. Figure 11 shows the confusion matrix. The overall accuracy of ResNet was 58.25%, and the calculation time was 55 min. Figure 12 displays the confusion matrix. In the comparison of A1-7 and ResNet, A1-7 was superior to ResNet in both overall accuracy and calculation time, as in the 10 classes.

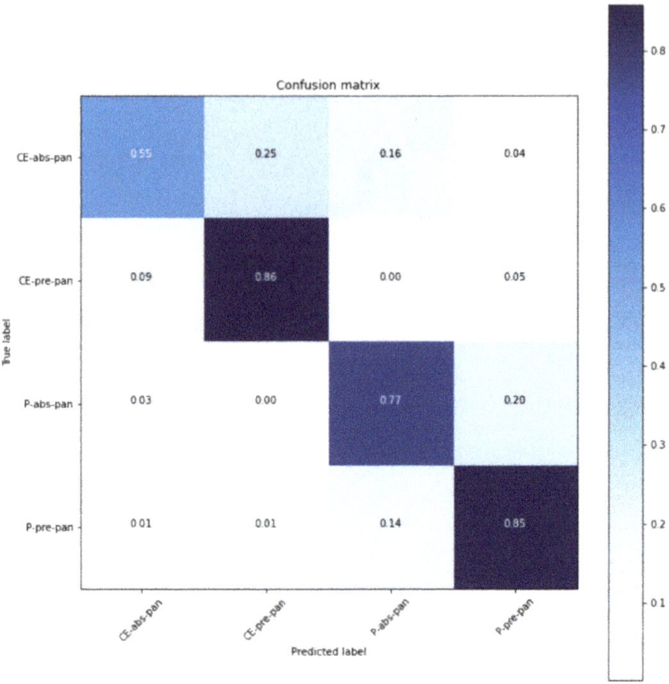

Figure 11. Confusion matrix of A1-7. (CE: contrast-enhanced, P: plain).

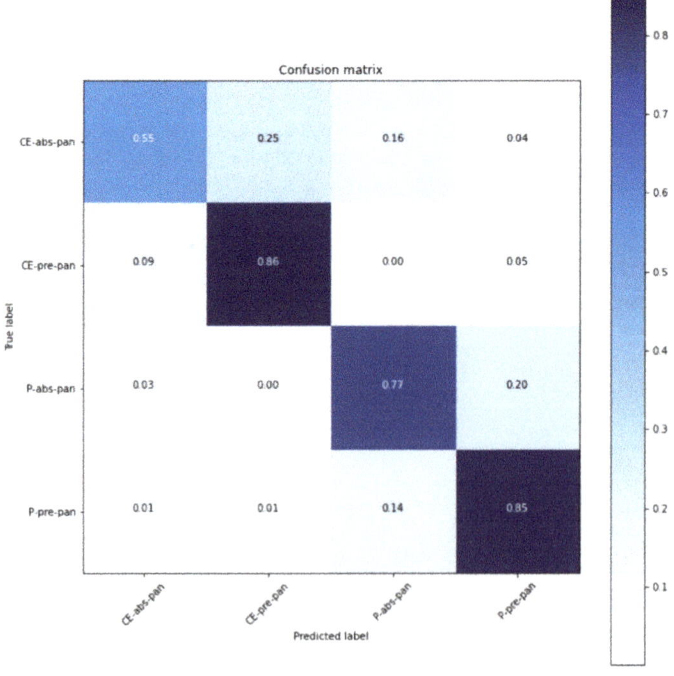

Figure 12. Confusion matrix of ResNet (CE: contrast-enhanced, P: plain).

4. Discussion

In the comparison between original (input image size: 224 × 224) and original (input image size: 512 × 512), original (input image size: 512 × 512) was found to have higher overall accuracy and a shorter calculation time. This means that training with large pixel-sized images may improve the overall accuracy and increasing the pixel size does not always increase the calculation time.

As a result of changing the filter size of the convolution layer, the overall accuracy was improved when the filter size was slightly increased, and it was decreased when the filter size was further increased in groups A to E. This result suggests that when training with images whose pixel size is larger than the original CNN, the overall accuracy is improved by appropriately increasing the filter size of the convolution layer. In the 512 × 512 image, the original filter size has a narrow range for the extraction of features, which makes it difficult to extract the overall features. Therefore, increasing the filter size improved the overall accuracy. However, if the filter sizes are made too large, detailed features cannot be extracted, and the overall accuracy is decreased. On the other hand, the calculation time became longer as the filter size increased. This is because the filter size was increased without changing the stride. The number of feature extractions was the same, but the range increased; thus, the calculation time became longer.

When the filter sizes of max-pooling were changed to an odd number from 5 to 15, the models with filter sizes 5, 7, and 9 exceeded original (input image size: 512 × 512), and the models after 11 were less than original (input image size: 512 × 512). This is because the overall features can be extracted by increasing the filter sizes of max-pooling to fit the 512 × 512 images. However, if the filter sizes of max-pooling are made too large, extraction of detailed features is not possible, and the overall accuracy is reduced. On the other hand, the calculation time was almost constant. Max-pooling is an operation used for the extraction of the maximum value, and the amount of the calculation is not large. Therefore, the calculation time was constant, even if the filter size was increased. When combining the filter sizes of the convolution layer and max-pooling, which had high overall accuracy, the overall accuracy of A1-7 was the highest. A1 and the model in which the filter size of max-pooling was 7 were not the most accurate models when changed separately. Combining the parameters of the model with the highest overall accuracy does not mean that a model with higher overall accuracy can be created. Thus, the overall accuracy varies depending on the affinity of the parameter combinations.

In the comparison of A1-7 and ResNet, A1-7 was superior in both overall accuracy and calculation time. According to the confusion matrix, ResNet often misclassified heads without contrast media as heads with contrast media. In contrast, the accuracy of A1-7 was 96%, which is about three times higher. Older CNNs, such as AlexNet, can exceed the overall accuracy of the relatively new ResNet by specializing the parameters to 512 × 512 images. Therefore, by specializing the parameters to the uncompressed pixel size, the overall accuracy of CNNs trained with compressed images might be improved.

Regarding the pancreas classification, A1-7 was superior to ResNet in both overall accuracy and calculation time, as in the 10 classes. According to the confusion matrix, the classification of images with the pancreas had about the same accuracy, but the accuracy of A1-7 in the classification of images without the pancreas was about twice that of ResNet. For this reason, A1-7 is not a CNN specialized for only 10 classes but is rather a generalized CNN.

In the study by Lakhani et al. [13], the authors used AlexNet and GoogLeNet [14] and created four types of image classification models with and without transfer learning, and the accuracy of the classification of tuberculosis was compared. Although the creation of multiple models and the accuracy comparison were similar to those in this study, Lakhani et al. did not change the CNN parameters, such as the filter size of the convolution layer. AlexNet, GoogLeNet, and transfer learning are technologies developed for general images and are not specialized for medical imaging. Therefore, in this study, we improved the accuracy by specializing the filter sizes of the convolution layer and max-pooling for medical images.

This study has five limitations. First, the comparison target was only ResNet. The created CNN was compared with ResNet, which is a typical example of high-performance CNN; however, the latest CNN is more accurate [15–20], so the created CNN should have been compared with the latest CNN. Second, we did not evaluate some parameters. However, considering that there is a limit to the number of parameters that can be evaluated individually and that we obtained a result of 94.40%, we believe that the number of parameters was sufficient. Third, we used the holdout validation. Ideally, k-fold cross-validation should be used, but we chose the holdout validation study because there were too many models to validate. In general, the ratio of data sets for holdout validation is 80:20 [5,21], and the mean performance is obtained from multiple data sets. In this study, the ratio of data sets was not 80:20, and the number of data sets was one. However, since some papers use a 90:10 ratio of data sets [22,23] and others use various ratios [1,6,24], this is not considered a problem. As for the number of data sets, it is not a problem since there are papers that validate with only one [3,4,25]. Fourth, only one generalization capability test was used in this study. By classifying small lesions, it is possible to verify the ability to respond to minute changes and clinical practicality. Since the accuracy in actual diagnosis is not verified, the classification of lesions and malignancy needs to be verified before it is used in clinical practice. However, being able to train with the original pixel size without resizing is useful, because it may capture more minute features. Finally, we tested the generalization capability only on the model with the highest overall accuracy. Testing with other models could reveal the relationship between accuracy and parameters.

The CNN we created in this study can quickly create a model, even if it is trained with a large number of medical images. This feature has the potential to create an image classification model that can be updated daily by training with images taken on the same day. As a result, the created CNN can be optimized for the imaging method, rules, radiologist habits, and patient tendency for each facility and can contribute to the creation of a diagnostic support application specialized for each facility.

5. Conclusions

By optimizing the filter size of the convolution layer and max-pooling of 512×512 images, we were able to quickly obtain a highly accurate medical image classification model. This improved CNN can be useful for the classification of lesions and anatomies for related diagnostic aid applications.

Author Contributions: K.M. contributed to data analysis, algorithm construction, and the writing and editing of the article; Y.A. and T.Y. contributed to reviewing and editing the paper; H.S. proposed the idea and contributed to data acquisition, performed supervision, project administration, and reviewed and edited the paper. All authors have read and agreed to the published version of the manuscript.

Funding: This research received no external funding.

Institutional Review Board Statement: The study was conducted according to the guidelines of the Declaration of Helsinki, and approved by the Hokkaido University Hospital Ethics Committee.

Informed Consent Statement: Informed consent was obtained from all subjects involved in the study.

Data Availability Statement: The created models in this study are available on request from the corresponding author. However, the image datasets presented in this study are not publicly available due to ethical reasons, e.g., containing information that could compromise the privacy of research participants.

Conflicts of Interest: The authors declare that no conflicts of interest exist.

References

1. Chae, K.J.; Jin, G.Y.; Ko, S.B.; Wang, Y.; Zhang, H.; Choi, E.J.; Choi, H. Deep Learning for the Classification of Small ($\leq$2 cm) Pulmonary Nodules on CT Imaging: A Preliminary Study. *Acad. Radiol.* **2020**, *27*, e55–e63. [CrossRef] [PubMed]
2. Nishio, M.; Sugiyama, O.; Yakami, M.; Ueno, S.; Kubo, T.; Kuroda, T.; Togashi, K. Computer-aided diagnosis of lung nodule classification between benign nodule, primary lung cancer, and metastatic lung cancer at different image size using deep convolutional neural network with transfer learning. *PLoS ONE* **2018**, *13*, e0200721. [CrossRef] [PubMed]
3. Sugimori, H. Evaluating the overall accuracy of additional learning and automatic classification system for CT images. *Appl. Sci.* **2019**, *9*, 682. [CrossRef]
4. Sugimori, H. Classification of Computed Tomography Images in Different Slice Positions Using Deep Learning. *J. Healthc. Eng.* **2018**, *2018*, 1753480. [CrossRef] [PubMed]
5. Pranata, Y.D.; Wang, K.C.; Wang, J.C.; Idram, I.; Lai, J.Y.; Liu, J.W.; Hsieh, I.H. Deep learning and SURF for automated classification and detection of calcaneus fractures in CT images. *Comput. Methods Programs Biomed.* **2019**, *171*, 27–37. [CrossRef] [PubMed]
6. Aghnia Farda, N.; Lai, J.Y.; Wang, J.C.; Lee, P.Y.; Liu, J.W.; Hsieh, I.H. Sanders classification of calcaneal fractures in CT images with deep learning and differential data augmentation techniques. *Injury* **2020**. [CrossRef] [PubMed]
7. Krizhevsky, A.; Sutskever, I.; Hinton, G.E. ImageNet classification with deep convolutional neural networks. *Commun. ACM* **2017**, *60*, 84–90. [CrossRef]
8. He, K.; Zhang, X.; Ren, S.; Sun, J. Deep residual learning for image recognition. In Proceedings of the IEEE Computer Society Conference on Computer Vision and Pattern Recognition, Las Vegas, NV, USA, 27–30 June 2016; pp. 770–778.
9. Santin, M.; Brama, C.; Théro, H.; Ketheeswaran, E.; El-Karoui, I.; Bidault, F.; Gillet, R.; Gondim Teixeira, P.; Blum, A. Detecting abnormal thyroid cartilages on CT using deep learning. *Diagn. Interv. Imaging* **2019**, *100*, 251–257. [CrossRef] [PubMed]
10. Kamiya, K.; Ayatsuka, Y.; Kato, Y.; Fujimura, F.; Takahashi, M.; Shoji, N.; Mori, Y.; Miyata, K. Keratoconus detection using deep learning of colour-coded maps with anterior segment optical coherence tomography: A diagnostic accuracy study. *BMJ Open* **2019**, *9*, 1–7. [CrossRef] [PubMed]
11. Nasrullah, N.; Sang, J.; Alam, M.S.; Mateen, M.; Cai, B.; Hu, H. Automated lung nodule detection and classification using deep learning combined with multiple strategies. *Sensors* **2019**, *19*, 3722. [CrossRef] [PubMed]
12. Roth, H.R.; Lu, L.; Farag, A.; Shin, H.C.; Liu, J.; Turkbey, E.B.; Summers, R.M. Deeporgan: Multi-level deep convolutional networks for automated pancreas segmentation. *Lect. Notes Comput. Sci.* **2015**, *9349*, 556–564. [CrossRef]
13. Lakhani, P.; Sundaram, B. Deep learning at chest radiography: Automated classification of pulmonary tuberculosis by using convolutional neural networks. *Radiology* **2017**, *284*, 574–582. [CrossRef] [PubMed]
14. Szegedy, C.; Liu, W.; Jia, Y.; Sermanet, P.; Reed, S.; Anguelov, D.; Erhan, D.; Vanhoucke, V.; Rabinovich, A. Going deeper with convolutions. In Proceedings of the IEEE Computer Society Conference on Computer Vision and Pattern Recognition, Boston, MA, USA, 7–12 June 2015; pp. 1–9.
15. Xie, Q.; Luong, M.T.; Hovy, E.; Le, Q.V. Self-training with noisy student improves imagenet classification. In Proceedings of the IEEE Computer Society Conference on Computer Vision and Pattern Recognition, Seattle, WA, USA, 13–19 June 2020; pp. 10684–10695.
16. Kolesnikov, A.; Beyer, L.; Zhai, X.; Puigcerver, J.; Yung, J.; Gelly, S.; Houlsby, N. Big Transfer (BiT): General Visual Representation Learning. *arXiv* **2019**, arXiv:1912.11370.
17. Xie, C.; Tan, M.; Gong, B.; Wang, J.; Yuille, A.L.; Le, Q.V. Adversarial examples improve image recognition. In Proceedings of the IEEE Computer Society Conference on Computer Vision and Pattern Recognition, Seattle, WA, USA, 13–19 June 2020; pp. 816–825. [CrossRef]
18. Zoph, B.; Vasudevan, V.; Shlens, J.; Le, Q.V. Learning Transferable Architectures for Scalable Image Recognition. In Proceedings of the IEEE Computer Society Conference on Computer Vision and Pattern Recognition, Salt Lake City, UT, USA, 18–23 June 2018; pp. 8697–8710.
19. Touvron, H.; Vedaldi, A.; Douze, M.; Jégou, H. Fixing the train-test resolution discrepancy. *arXiv* **2019**, arXiv:1906.06423.
20. Liu, C.; Zoph, B.; Neumann, M.; Shlens, J.; Hua, W.; Li, L.J.; Fei-Fei, L.; Yuille, A.; Huang, J.; Murphy, K. Progressive Neural Architecture Search. *Lect. Notes Comput. Sci.* **2018**, *11205 LNCS*, 19–35. [CrossRef]
21. Wang, G.; Sun, Y.; Wang, J. Automatic Image-Based Plant Disease Severity Estimation Using Deep Learning. *Comput. Intell. Neurosci.* **2017**, *2017*. [CrossRef] [PubMed]
22. Ding, Y.; Sohn, J.H.; Kawczynski, M.G.; Trivedi, H.; Harnish, R.; Jenkins, N.W.; Lituiev, D.; Copeland, T.P.; Aboian, M.S.; Aparici, C.M.; et al. A deep learning model to predict a diagnosis of Alzheimer disease by using 18 F-FDG PET of the brain. *Radiology* **2019**, *290*, 456–464. [CrossRef] [PubMed]
23. Zhou, Q.Q.; Wang, J.; Tang, W.; Hu, Z.C.; Xia, Z.Y.; Li, X.S.; Zhang, R.; Yin, X.; Zhang, B.; Zhang, H. Automatic detection and classification of rib fractures on thoracic ct using convolutional neural network: Accuracy and feasibility. *Korean J. Radiol.* **2020**, *21*, 869–879. [CrossRef] [PubMed]
24. Chilamkurthy, S.; Ghosh, R.; Tanamala, S.; Biviji, M.; Campeau, N.G.; Venugopal, V.K.; Mahajan, V.; Rao, P.; Warier, P. Deep learning algorithms for detection of critical findings in head CT scans: A retrospective study. *Lancet* **2018**, *392*, 2388–2396. [CrossRef]
25. Tomita, N.; Cheung, Y.Y.; Hassanpour, S. Deep neural networks for automatic detection of osteoporotic vertebral fractures on CT scans. *Comput. Biol. Med.* **2018**, *98*, 8–15. [CrossRef] [PubMed]

 applied sciences

MDPI

Article

A Provable and Secure Patient Electronic Health Record Fair Exchange Scheme for Health Information Systems

Ming-Te Chen † and Tsung-Hung Lin *,†

Department of Computer Science and Information Engineering, National Chin-Yi University of Technology, Taichung 41170, Taiwan; mtchen@ncut.edu.tw
* Correspondence: duke@ncut.edu.tw
† These authors contributed equally to this work.

Abstract: In recent years, several hospitals have begun using health information systems to maintain electronic health records (EHRs) for each patient. Traditionally, when a patient visits a new hospital for the first time, the hospital's help desk asks them to fill in relevant personal information on a piece of paper and verifies their identity on the spot. This patient will find that many of her personal electronic records are in many hospital's health information systems that she visited in the past, and each EHR in these hospital's information systems cannot be accessed or shared between these hospitals. This is inconvenient because this patient will again have to provide their personal information. This is time-consuming and not practical. Therefore, in this paper, we propose a practical and provable patient EHR fair exchange scheme for each patient. In this scheme, each patient can securely delegate the information system of a current hospital to a hospital certification authority (HCA) to apply migration evidence that can be used to transfer their EHR to another hospital. The delegated system can also establish a session key with other hospital systems for later data transmission, and each patient can protect their anonymity with the help of the HCA. Additionally, we also provide formal security proofs for forward secrecy and functional comparisons with other schemes.

Keywords: electronic health records; fair exchange; forward secrecy

Citation: Chen, M.-T.; Lin, T.-H. A Provable and Secure Patient Electronic Health Record Fair Exchange Scheme for Health Information Systems. *Appl. Sci.* **2021**, *11*, 2401. https://doi.org/10.3390/app11052401

Academic Editor: Federico Divina

Received: 25 January 2021
Accepted: 2 March 2021
Published: 8 March 2021

Publisher's Note: MDPI stays neutral with regard to jurisdictional claims in published maps and institutional affiliations.

Copyright: © 2021 by the authors. Licensee MDPI, Basel, Switzerland. This article is an open access article distributed under the terms and conditions of the Creative Commons Attribution (CC BY) license (https://creativecommons.org/licenses/by/4.0/).

1. Introduction

In recent years, many research topics have arisen to make human life more convenient. An electronic health record (EHR) is an integrated personal medical record in health information systems. Many countries implement their own health information systems to help manage each patient's activities and keep track of their health. We can imagine a scenario in which a patient (let us call her Alice) plans to go to a new hospital and sees a doctor. In this situation, she may have to fill in her personal medical information another time when she attends a new hospital. In addition, if her doctor needs to know her medical treatment history from other hospitals, how she provides these records securely to her doctor needs to be considered. These problems are especially urgent. Our proposed scheme ensures the ease and security of data access and migration. Our approach proposes a practical and provable patient EHR fair exchange scheme with session key establishment for health information systems. Patients cannot only delegate the migration of their personal EHR to a desired hospital system from their current hospital health information system but also protect their privacy. Our mechanism provides secure data storage and the secure transfer of authorized information to a designated location. This study has two limitations. First, we assume that each patient's EHR record is well defined and appropriate for each healthcare facility. The process of electronic health information record transmission at each hospital provider is easily done by implementing our proposed scheme for secure encrypted transmission without considering issues such as different forms or file names or a lack of formatting details. Second, each facility will transfer or link the patient's EHR to other facilities through patient consent or under a national policy when the patient requires better care at those facilities.

Summarizing all problems, we propose a high-level practical and provable patient EHR fair exchange scheme with key agreement for health information systems. Not only could a patient delegate the health information systems of the current hospital to migrate their personal EHR to the desired hospital system, but they can also keep their privacy. Our mechanism provides data storage and the secure transfer of authorized information to designated locations. What information can be authorized is beyond the scope of this study to determine. For example, whether COVID-19 patient privacy concerning patients' names, identities, and genetic sequences can be transmitted to different hospitals is beyond this study's scope. The mechanism presented here could guarantee data transfer and storage safely and securely. What is more, our scheme also provides a formal security proof in a random oracle model under chosen-ciphertext security.

This paper is organized as follows. Section 2 introduces related works. Section 3 deploys security definitions. Section 4 shows our proposed method. Section 5 describes our security analysis. Section 6 provides a security proof. Finally, Section 7 presents our conclusions.

2. Related Works

In this section, we surveyed some articles [1–4]. In [1], the author only mentioned how EHRs are used and managed. The author also talked about the EHR format that followed the definition of HL7 [5] and that performed well-known protocols to encode each patient's EHR from TCP/IP, MIME, HTTP(S), and SOAP.

In [2], the authors discussed several security requirements, such as EHR storage security, malicious code prevention, protected access right management, and other aspects to protect the health information system. However, they did not provide a practical scheme that would allow a patient to migrate their EHR to a health information system. We can imagine a scenario where a hospital only adopts the above simple protocols to develop its own health information system without any security mechanism. Additionally, it is not feasible for each patient to perform their own EHR exchange under this scheme.

On the other hand, in [5], the authors suggested that each patient's health records (or files) could be portably stored on a flash disk. This idea is appealing but is currently difficult to implement. There are many security issues to be handled, including portable device security and patient medical file access rights. However, more security mechanisms are needed to solve these kinds of security issues, which are beyond the scope of this research.

In addition, various patient authentication schemes of e-health systems have been proposed [6–10]. In [6,7], the schemes suffered a user impersonation attack and did not offer session key establishment with a formal security proof. The authors in [8–10] did not provide session key establishment with a forward secrecy proof. In [11,12], the authors each proposed a framework with a patient-centric access right in a blockchain environment. However, they did not provide a practical mechanism for each patient to perform EHR migration exchange securely.

Additionally, many studies are now examining the importance of personal privacy and data authorization. For example, the prevalence of COVID-19 has made many patients reluctant to disclose information about their infection, but government healthcare units want to control the trajectory of tracking these patients. A method of providing improvements in these mechanisms is the main motivation and purpose of our study. Therefore, in this paper, we emphasize providing a secure, simple, and complete mechanism for authorizing data transfer during personal information migration and demonstrate that our approach is secure and effective in practice through a professional information security authentication model.

Hence, we summarize and list here seven kinds of security attack when a patient attempts to migrate their personal information data through a traditional authentication model:

- Replay Attack Resistance: A malicious attacker intercepts the parameters used in the mutual authentication transaction successfully. They can then forward these parameters again to impersonate one party to communicate with other parties during the mutual authentication transaction and vice versa.
- Resist User Impersonation: A malicious attacker impersonates some party by replaying the intercepted signatures or random variables to other parties engaged in the mutual authentication transaction.
- Mutual Authentication: Without mutual authentication with other e-health systems, an attacker can pretend as a fake system to let other patients register and login. Then, patient's EHRs information cannot be stored securely in this storage location.
- Data Security Problem: An attacker intercepts the parameters, including ciphertext, successfully during the mutual authentication transaction, and they may then decrypt the intercepted ciphertext by adopting these intercepted parameters.
- Session Key Establishment with Forward Secrecy: Each party communicates a temporary symmetric key for data transmission after performing mutual authentication successfully. If the session key is easy to guess or derive by an attacker successfully without forward secrecy, then this attacker can derive the used session keys before the next mutual authentication phase.
- EHR Fair Exchange Problem: If there is packet loss or data loss during the EHR migration transaction, a patient's EHR can be lost when they attended a desired hospital. At this time, without their personal EHR, the e-health system of this hospital can delay their medical treatment in this situation.
- Patient's Anonymity Protection: During the EHR migration transaction, if a patient's EHR identity information is not protected well, then it could be exposed and intercepted by an attacker. Additionally, the patient's EHR information may be misused by an attacker further.

Our contribution is to offer an efficient provable and practical patient EHR fair exchange scheme so that each patient can migrate their personal EHR securely from one hospital to another and provides solutions to the above seven problems. We designed a secure patient EHR exchange protocol that can be integrated into the e-health information system of each hospital. The proposed scheme could also guarantee convenience, rapidity, and integrity. We constructed a high-level practical and provable patient EHR fair exchange scheme with key agreement for the health information system. A patient could not only delegate the current hospital's health information systems to migrate their personal EHR to the desired hospital system, but also keep their privacy. Additionally, our scheme demonstrates a formal security proof with light-weight computation for both authentication parties.

3. The Proposed Scheme

Our proposed scheme contains three stages: the migration registration phase, the EHR migration phase, and the data recovery phase.

3.1. Preliminary

In this subsection, we provide some definitions in our proposed scheme.

- n: A large prime number that forms a finite primes field with an order less than n.
- l: A security parameter that defines the hashed messages' length.
- V: The current medical organization.
- W: The patient's desired medical organization.
- U: A patient making an authentication request with a server V in a health information system.
- S: A server that accepts the registration of the patient request, the login request, and the password modification request in the health information system of hospital V.
- UID_i: A patient's real identity computed from social security numbers, where $i \in \{U\}$ in the certification.

- ID_i: A registration local identity that can link the UID_i of user $i \in \{U\}$.
- H_1, H_2: Two secure hash functions that each maps $Z_n^* \to \{0,1\}^l$ with collision-resistance and outputs the same l-bits hash strings.
- pw_U: A initial password that is chosen by a server V when a patient U has remotely registered on the server for the first time.
- E_{k_i}: A symmetric key encryption function for the party i under the symmetric key k_i, where $i \in \{U, V\}$.
- D_{k_i}: A symmetric key decryption function for the party i under the symmetric key k_i, where $i \in \{U, V\}$.
- ASE_{pk_i}: An asymmetric key encryption function for the party i, where $i \in \{U, V\}$.
- ASD_{sk_i}: An asymmetric key decryption function for the party i, where $i \in \{U, V\}$.
- EHR_i: A patient electronic health record (EHR) in one hospital organization, where $i \in U$.
- Bio_i: A biometric information value that is chosen uniformly from the party i, where $i \in \{U\}$.
- $Date$: A patient EHR migration limitation date period.
- $Cert_i$: A migration certification of the party i with ID_i registration and public keys for this ID_i, where $i \in \{U, V\}$.
- HCA: A hospital certification authority (HCA) that helps the patient to generate the patient migration permission signature to another hospital or medical center in the public key infrastructure (PKI).
- $Agree_{i->j}$: A patient-delegated EHR migration agreement document whereby the patient agrees that its own patient files can migrate from current hospital i to the desired one j, where $i, j \in \{V, W\}$.

3.2. The Migration Registration Phase

Before starting this phase, a patient (U) forwards (UID_U, ID_U) to a hospital certification authority (HCA) for migration certification registration with a secure channel. After receiving this identity (UID_U, ID_U), the HCA keeps this link information and generates $Cert_i$ certification with ID_i for EHR migration and forwards this $Cert_i$ to the patient U.

When U performs this phase with the server V of the current hospital, U forwards a patient migration registration request to a server (V). After receiving this request, the server V forwards this request to the HCA to help U obtain the permission signature from HCA. V first prepares two hash functions: one is H_1, and the other is H_2, where $H_1 : Z_n^* \to \{0,1\}^l$ and $H_2 : Z_n^* \to \{0,1\}^l$.

- First, U has prepared their biometric value Bio_U and computed a random value r_{B_U} with a random number $r''_U \in Z_n^*$, where $r_{B_U} = r''_U \oplus H_1(Bio_U)$. After the above is computed successfully, they forward their identity ID_U, which is computed from $H_1(r''_U||UID_U)$ and encrypted via real identity cipher-text $C_{HCA} = AE_{pk_{HCA}}(r''_U, Bio_U, UID_U, ID_U, Cert_U, EHR_U \oplus r''_U)$ with public key pk_U and their certificate $cert_U$ with their biometric information r_{B_U}, to the server V. It also generates the applicant-delegated migration signature S_U, where the signature S_U is to be the $Sig_U(Agree_{V->W}, UID_U, ID_U, EHR_U \oplus r''_U)$ with the registration identity ID_U, real identity UID_U, and EHR migration delegated agreement $Agree_{V->W}$ with final ciper-text $EHR_U \oplus r''_U$. U then prepares the random number r''_V, where $r''_V = r_V \oplus r_{B_U}$, and forwards these files to the server V with $(r''_V, C_{HCA}, S_U, Cert_U, Agree_{V->W})$.
- After V receives these messages, it can check the S_U with the above messages and forward the C_{HCA} to the HCA. When HCA has received this message with S_U, it can decrypt C_{HCA} to obtain all parameters. First, it can fetch the random value $r''_V \oplus r_{B_U} = r_V$ and verify the signature S_U with other parameters. If they are valid, it saves these files for data recovery usage. It then generates S_{HCA}, which is $Sig_{HCA}(S_U, Date, ID_U)$. Finally, it returns $(S_{HCA}, H_1((r_V + 1)||r_{HCA}), (r_V + 1) \oplus r_{HCA}, Date, ID_U, S_U)$ to the server V.

- V receives this message tuple, where one is $(S_{HCA}, H_1((r_V+1)|r_{HCA}))$ and the other is $((r_V+1) \oplus r_{HCA}, Date, ID_U, S_U)$, and it can verify the signature S_{HCA} with the above parameters and compute $r_{HCA} = (r_V+1) \oplus (r_V+1) \oplus r_{HCA}$. When the above messages are valid, V returns $H_1(r_{HCA}+1)$ and forwards it back to the server HCA. In addition, it also generates a signature $Sig_V(S_{HCA}, S_U, Agree_{V->W})$ as the receipt S_V and finishes this phase after forwarding S_V and S_{HCA}. We demonstrate in the Figure 1.

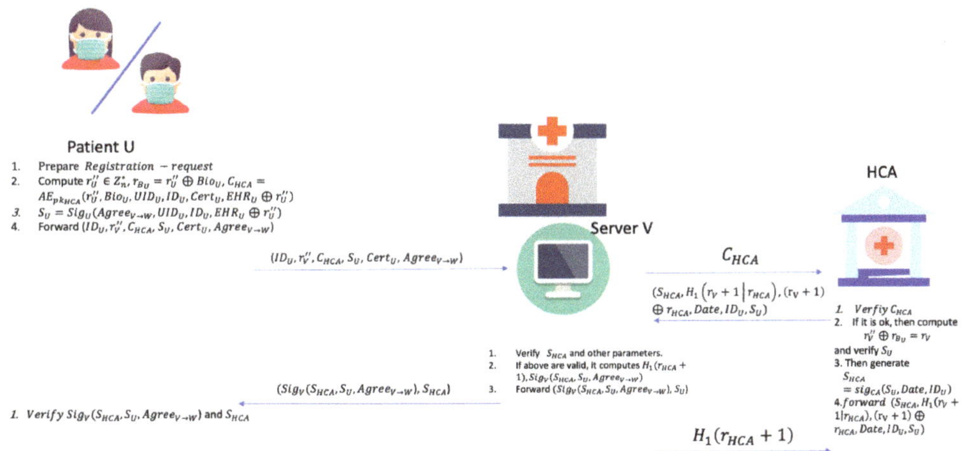

Figure 1. The migration registration phase.

3.3. The EHR Migration Phase

In this phase, the server V will behave according to the delegated agreement file $Agree_{V->W}$ with the signature S_U, and it prepares these messages as follows.

- First, it selects a random r^*_V and computes $C_V = AE_{pk_W}(r^*_V)$. It then forms the message tuple $(C_V, S_U, S_{HCA}, EHR_U \oplus r''_U, Date)$ for the server of the desired migration hospital W. When the system of W has received this migration agreement from V, it verifies this message tuple and decrypts C_V to obtain the challenge random number. If all signatures and parameters are valid, it generates the response random number $r^*_W \oplus H_2(r^*_V+1)$ and returns it to the server V.
- When V receives $r^*_W \oplus H_2(r^*_V+1)$, it decrypts it with r^*_V+1. If it is valid, it can obtain r^*_W and computes the response random number $H_2((r^*_W+1) \oplus r''_U), (r^*_W+1) \oplus r''_U$ for the server of hospital W. Finally, W receives $H_2(r^*_W+1 \oplus r''_U), (r^*_W+1) \oplus r''_U$ from V. It also verifies the message to check if it was returned by the real V. If true, it decrypts $(r^*_W+1) \oplus r''_U$ to obtain r''_U, computes the session key $ssk_{V,W} = H_1((r^*_W+1)||(r^*_V+1))$, and decrypts $EHR_U \oplus r''_U$ to fetch the patient's EHR file EHR_U with the random r''_U.
- After performing mutual authentication successfully, V can also use the session key $ssk_{V,W}$ to generate an cihper-text, such as $E_{ssk_{V,W}}(EHR_U)$, for the rest of the data transmission of the patient U's EHR. Figure 2 shows the scenario of this phase.

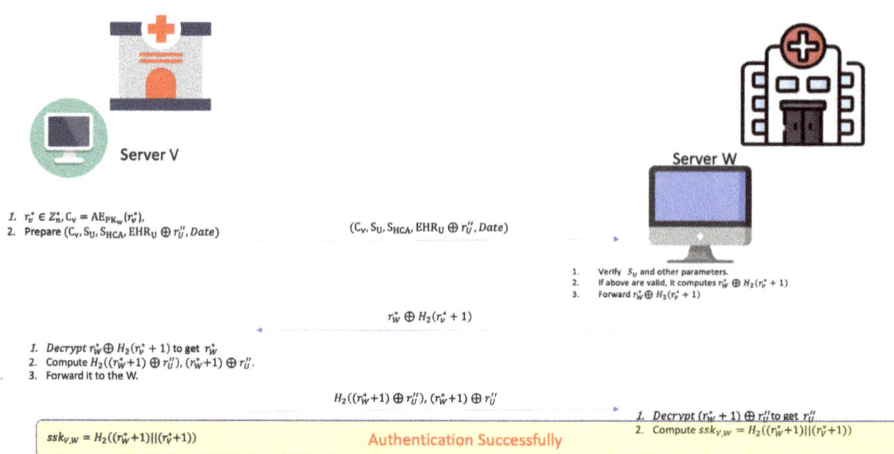

Figure 2. The EHR migration phase.

3.4. The Data Recovery Phase

In this phase, if there is some network packet loss or EHR data loss of the patient U after they have performed the EHR migration phase, then the e-health system of the hospital W cannot obtain the full U's personal EHR, so U can ask the HCA to deal with this situation.

- First, when a patient migrating to a hospital W that does not receive the EHR_U from the server V after querying the system of hospital W, then U can ask HCA to resolve the situation with some evidence $Sig_V(S_{HCA}, S_U, Agree_{V->W})$ with signatures S_{HCA}, S_U and $Agree_{V->W}$. After verifying the above signatures successfully, HCA can fetch the $EHR_U \oplus r''_U$ to the server of hospital W with $AE_{pk_W}(r''_U)$.
- W can then decrypt $AE_{pk_W}(r''_U)$ to obtain r''_U and fetches the patient's EHR file EHR_U from the above $EHR_U \oplus r''_U$ by applying $\oplus$ with r''_U. At this time, this situation is solved with HCA's help if needed. This phase's scenario shows in the Figure 3.

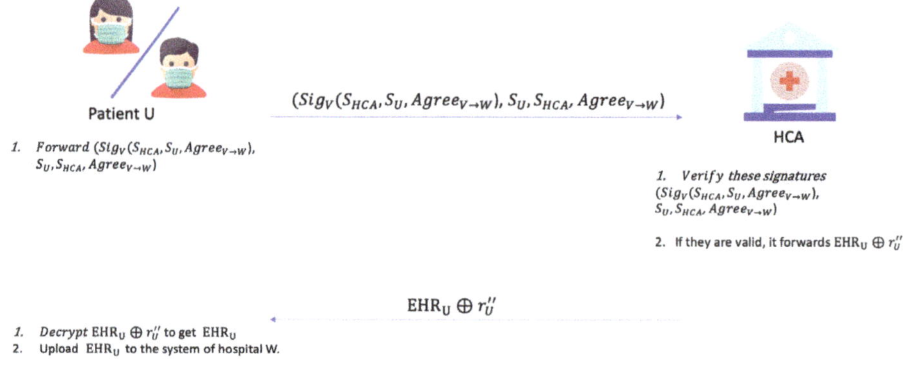

Figure 3. The data recovery phase.

4. Security Assumptions

4.1. Secure Digital Signature

In this scheme, we define a secure digital signature. In the beginning, we have that $Sig(\cdot)$ is a signature generation function that inputs a message m with a signer's secret key sk_i and outputs a signature S_i. We also assert this signature function is based on the RSA factoring hard problem or the discrete logarithm problem. We can then input a signature such as S_i with the signer's public key pk_i into the verification function $Ver(\cdot)$ and see what the output is. If the output is 1, then we can confirm the signature S_i is valid and signed by the signer i. In this scheme, we also assumed that the signature building block is under the RSA problem. If there is an attacker, we assume it as $\mathcal{F}^*$. If $\mathcal{F}^*$ can make a forged $l+1$ signature called $S'_{i,j+1}$ of some user $i \in \{U, V, HCA\}$ in at most $l$signature queries, and this signature can pass the verification $Ver(pk_i, S'_{i,j+1})$ successfully with non-negligible probability ε, then $\mathcal{F}^*$ can be used to break the RSA factoring problem. Thus,

$$Pr[S'_{i,j+1} \longleftarrow \mathcal{F}^*(Sig(sk_i,\cdot), Ver(pk_i,\cdot), i \in \{U, V, HCA\}) | Ver(S'_{i,j+1}) = 1] \geq \varepsilon. \quad (1)$$

4.2. Unforgeability

In this scheme, we define the secure digital signature scheme in the above. First, we define an attacker $\mathcal{F}^*$, whose ability is to forge a signature that can be verified successfully through the $Ver(\cdot)$ verification function with non-negligible probability ε'. We also define a simulator $\mathcal{D}$ that adopts $\mathcal{F}^*$'s ability to break the underlying hard problem (such as the RSA factoring problem) in the above secure signature scheme. After $\mathcal{D}$ is given the environment parameters $G(\cdot)$, it can start the protocol simulation with $\mathcal{F}^*$. $\mathcal{F}^*$ can make the signature queries to the $\mathcal{D}$. $\mathcal{D}$ will also output the signature back according to the received input m from $\mathcal{F}^*$ on some user i. After this simulation, if $\mathcal{F}^*$ generates a forged signature $S'_{i,j+1}$, the verification result of $S'_{i,j+1}$ is valid. We then have

$$\begin{aligned} Pr[\mathcal{D}^{\mathcal{F}^* \longrightarrow S'_{i,j+1}} | \text{Use } S'_{i,j+1} \text{ to solve the RSA factoring problem}] \geq \\ Pr[S'_{i,j+1} \longleftarrow \mathcal{F}^*(Sig(sk_i,\cdot), Ver(pk_i,\cdot), i \in \{U, V, HCA\}) | Ver(pk_i, S'_{i,j+1}) = 1] \geq \varepsilon'. \end{aligned} \quad (2)$$

In fact, if there is no attack $\mathcal{F}^*$ that can make a forged signature pass the verification successfully with non-negligible probability ε, then we cannot use $\mathcal{F}^*$ to solve the RSA factoring problem with non-negligible ε' probability.

Lemma 1 (Unforgeability). *First, we define Sig, which is a secure digital signature function and equips two secure hash functions, H_1 and H_2, which can be replaced with two random oracles functions RO_1 and RO_2. In our proposed EHR scheme, we also define our proposed EHR scheme with unforgeability (Unf), which satisfies the following situations. In other words, if Sig is (t', ε') and unforgeable, then*

$$Adv^{Unf}_{\mathcal{F}^*, Sig^{H_1, H_2, RO_1, RO_2}}(\theta, t') \leq \frac{1}{2 \cdot I^3 \cdot q_s} + \varepsilon', \quad (3)$$

where t' is the maximum total experiment time, including an adversary execution time, I is an upper bound on the number of parties, at most signature oracle q_s, and ε' is taken over the coin flip of our EHR scheme.

4.3. Indistinguishability

We define an attacker $\mathcal{A}$ on the experiment **EXP** of our symmetric encryption/decryption functions (**SE**), which is a game controlled by the simulator $\mathcal{S}$. We also define two pseudo-random hash functions (ω_1 and ω_2), which are satisfied with the property we call "indistinguishability" (**Ind**), due to which the attacker $\mathcal{A}$ can make a hash query to ω_1 and ω_2 on the message M', which is chosen by $\mathcal{A}$. These functions act as real functions as our hash functions (H_1 and H_2), where $i \in \{U, V\}$. The simulator also can switch this function pair to respond to each query made by $\mathcal{A}$ during the simulation rounds of the above experiment. Finally, the simulator $\mathcal{S}$ is given a challenge message target M chosen by the $\mathcal{A}$.

At this time, $\mathcal{S}$ makes a coin flip on b. If $b = 0$, $\mathcal{S}$ randomly chooses (ω_1, ω_2) to generate the hashed value of M and return it to $\mathcal{A}$. Otherwise, $\mathcal{S}$ forwards M to (H_1, H_2) to ask for the hash value. $\mathcal{A}$'s goal is to guess correctly the hashed value that is from (ω_1, ω_2) or (H_1, H_2) with non-negligibility probability.

Lemma 2 (Indistinguishability). *In this lemma, our symmetric encryption/decryption functions satisfy the indistinguishably property if there is no attacker $\mathcal{A}$ that can guess the hashed value from the chosen (M) with more than $\frac{1}{2}$ with negligible probability ε^* under the t^* polynomial time bound. That is,*

$$|Pr[b' \longleftarrow \mathcal{F}^{(\omega_1,\omega_2,H_1,H_2)}(M)|b = b'] - \frac{1}{2}| \leq \varepsilon.$$

Therefore, we concluded that

$$Adv^{Ind}_{\mathcal{A},SE}(\theta, t^*) \leq \frac{1}{2} + \varepsilon^*.$$

4.4. Indistinguishable-Chosen Cipher-Text Attack (Ind-CCA)

In this scheme, we define our proposed asymmetric encryption/decryption function (**ASE**), which satisfies the semantic security in the following definitions.

First, we define an attacker $\mathcal{A}$ that can ask encryption/decryption queries in our scheme, respectively. However, the attacker $\mathcal{A}$ can also make an encryption query to the chosen message that we define as M'. The attacker $\mathcal{A}$ can then also make a decryption query to the decryption oracle, whose task is to decrypt the cipher-text sent $\mathcal{A}$. Next, we define *Game*, which is the simulation of our proposed scheme that can equip many different oracles, and oracles can answer back to the adversary depending on the attacker's input messages. We also define some oracles, such as the encryption oracle $AE_{pk_T}(\cdot, \theta)$ with the security parameter θ. This encryption oracle can generate the ciphertext according to the received input M_b, where $b \in \{0, 1\}$. In addition, we also model the decryption oracle that receives the cipher-text C from the attacker $\mathcal{A}$ and returns the final decrypted message M to the attacker $\mathcal{A}$. In the following, we consider two situations involving $\mathcal{A}$.

Phase 1: In this phase, the attacker $\mathcal{A}$ can make the decryption and encryption queries on a chosen message (call it M'). I.e., if $\mathcal{A}$ makes an encryption query on the input message M', then $C' \longleftarrow AE_{pk_T}(M', \theta)$ returns to $\mathcal{A}$. At this time, $\mathcal{A}$ can also make the decryption query on cipher-text C', and the simulator will then forward this C' to the decryption oracle and return the final message M' back to the $\mathcal{A}$. Additionally, $\mathcal{A}$ can also make other kinds of queries, such as a hash query to the hash oracles.

Challenge: In this phase, if $\mathcal{A}$ has performed training on the above encryption/decryption query many times, then, in the following challenge phase, the attacker $\mathcal{A}$ will choose a challenge message pair (M_0^*, M_1^*) for the simulator for game playing. The simulator then will toss the coin on b after it receives this message pair. If the final output b is 1, then we can have $C^* \longleftarrow AE_{pk_T}(M_b^*, \theta)$. Otherwise, we have $C^* \longleftarrow AE_{pk_T}(M_{1-b}^*, \theta)$. After the attacker $\mathcal{A}$ has asked the cipher-text on the chosen target messages (M_0^*, M_1^*), the only restriction is that the $\mathcal{A}$ cannot ask the decryption oracle on the target message (M_0^*, M_1^*) with the input cipher-text C^*. This query can make the simulation fail due to the simulator cannot be able to tell the answer of cipher-text C^*. Except in the above query, $\mathcal{A}$ can make other kinds of queries on different messages.

Lemma 3. *In this lemma, we model the above actions as the game simulation steps, which we played with the attacker $\mathcal{A}$.*

$Game^{Ind-CCA-b}_{\mathcal{A},ASE}(\theta)$
Phase 1.
$T \in \{U,V\}, \{M_0, M_1\} \longleftarrow \mathcal{A}^{ASE_{pk_T}(\cdot,\theta),ASD_{sk_T}(\cdot,\theta),H_1(\cdot),H_2(\cdot)}$
Challenge Phase.
$b \in \{0,1\}, C^* \longleftarrow ASE_{pk_T}(M_b^*, \theta)$,
$b' \longleftarrow \mathcal{A}^{ASE_{pk_T}(\cdot,\theta)}(C^*, M_0^*, M_1^*)$
Return b'.

The advantage function of the adversary that $\mathcal{A}^{Ind-CCA}_{ASE}(\cdot,\theta)$ is defined as $Adv^{Ind-CCA}_{\mathcal{A},ASE}(\theta) = |Pr[Game^{Ind-CCA-1}_{\mathcal{A},ASE}(\theta) = 1] - Pr[Game^{Ind-CCA-0}_{\mathcal{A},ASE}(\theta) = 1]| < \frac{1}{2}|Pr[Game^{Ind-CCA-1}_{\mathcal{A},ASE}(\theta) = 1]| \leq \varepsilon'$.

4.5. Partner Function

In this definition, we define the partner function. We assume that there is an instance Π_i^k whose action is the same as player i in the k-th session, where $i, j \in \{U,V\}$ and $k \in N$, where N is the number for total players. Let the partner function be the instance of player j (call it $\Pi_j^{k'}$) in the k'-th session, where $i, j \in \{U,V\}$ and $k' \in N$. At this time, the instances Π_i^k and $\Pi_j^{k'}$ believe that each side is the real player $i, j \in \{U,V\}$ in the $k, k' \in N$ session, respectively. At this time, we can say that two instances Π_i^k and $\Pi_j^{k'}$ are partnered if the following statements are true:

1. Π_i^k's session identity is the same as the session identity of $\Pi_j^{k'}$.
2. p_i is the partner of $\Pi_j^{k'}$ in the session k' of $\Pi_j^{k'}$.
3. p_j is the partner of Π_i^k in the session k of Π_i^k.

4.6. Freshness

In this definition, we define freshness. We assume that there is an instance where Π_i^k is "fresh" if it satisfies the following conditions.

1. Π_i^k has not been queried the reveal query $Reveal(i,k)$.
2. There is a partner $\Pi_j^{k'}$ that is matched to partner Π_i^k by the partner function, and Π_i^k has not been queried the reveal query $Reveal(j,k')$.
3. The partner of Π_i^k is not the inside attacker during communication in the instance of the player j.

4.7. Forward Secrecy (FS)

Our proposed two factor patient authentication scheme is forward secrecy (FS) if $\mathcal{A}$ cannot compromise the past information, even if they have sent $Corrupt(i)$ (or $Corrupt(j)$) to the player i, where $i, j \in \{U,V\}$.

Theorem 1. *First, we assume that ASE is an indistinguishable-CCA (Ind-CCA) secure asymmetric encryption/decryption scheme and equips two secure hash functions, H_1 and H_2, which we can be replaced with two random oracle (RO) functions, respectively. We also assume that our proposed patient electronic health record exchange scheme (PEHRES) that is forward secure (FS) and unforgeable (Unf) also satisfies the following situations. In other words, if our proposed scheme is secure, then*

$$
\begin{aligned}
Adv_{PEHRES}^{FS,Unf,Ind-CCA}(\theta,t) \leq \frac{1}{2}(I^2 q_h q_e q_s(Adv_{ASE,\mathcal{D},C^*_{HCA}}^{Ind-CCA}(\theta,t')) + 1) + \\
\frac{1}{2}(I^2 q_h q_e(Adv_{ASE,\mathcal{D},C^*_V}^{Ind-CCA}(\theta,t')) + 1) + \\
\frac{1}{2}((Iq_h)^2 Adv_{\mathcal{A},SE}^{Ind}(\theta,t^*) + 1) + (I^3 q_s) Adv_{Sig,\mathcal{S},\mathcal{F}}^{Unf}(\theta,t^*) + \varepsilon, t \leq t' + t^*,
\end{aligned} \tag{4}
$$

where t is the total execution time, t' is the maximum total experiment time including an adversary execution time, t^* is the maximum total time to guess the real session key, I is an upper bound on the number of parties, with at most q_e encryption queries at most q_s decryption oracles, and q_h is an upper bound on the number of H_1 and H_2 queries in the experiment, where ε is a negligible advantage.

5. Security Analysis

In this section, we provide security analysis and functional analysis of our proposed scheme.

5.1. Replay Attack Resistance

In this EHR migration phase, we adopt random values r''_U, r^*_V, and r^*_W as our authentication challenge numbers. We assume an attacker can capture authentication messages among the protocol communication and may replay these captured messages to the server W to impersonate the patient U. First, the server V will check that this message was used before in some session before communicating with the server W. Hence, the server V will also check that one of these messages r''_U, r^*_V, and r^*_W was used before. If one of them was used, then it would close this session and save the record as the replay attack from V.

5.2. Resist User Impersonation Attack

In this proposed scheme, the adversary cannot replay any authentication message without the user U's biometric information Bio_U, and it also cannot guess the random number r''_U successfully to impersonate the server V. Additionally, the adversary does not have the non-negligible probability to forge the patient's signature to the server V. In addition, the server V also checks the signature S_U to authenticate the patient U's identity in the migration registration phase. Thus, the adversary cannot have non-negligible probability to forge U's signature S_U under the RSA factoring problem. Therefore, our scheme can resist user impersonation attacks.

5.3. Provide Mutual Authentication

In the EHR migration phase, a patient U can delegate the server V to perform the EHR migration exchange with the system of the desired hospital W. Server V can perform the challenge response with the server of W, and they both communicate a session key for later usage after successful authentication. During the authentication rounds, V and W can check the freshness of random numbers (r''_U, r^*_V, and r^*_W). If one of them is to be replayed, V or W would find out and deny this session with the other party. Finally, it would close this phase and record that there was a replay attack in this EHR migration phase.

5.4. Provide Data Security

In the EHR migration phase, all random numbers are generated by these two parties and drop off when the authentication between them is successful. In addition, not only are r''_U, r^*_V, and r^*_W verified by these two parties V and W, but also they can also be response messages to confirm their respective identities.Hence, the adversary cannot have a non-negligible probability to replace each of these messages to pass the authentication process. In the data recovery phase, r''_U is used to encrypt the patient's EHR, and the adversary does not have a non-negligible probability to obtain a patient's EHR, under the assumption that the symmetric encryption/decryption function is indistinguishable for the adversary in a polynomial time bound.

5.5. Session Key Establishment

In the EHR migration phase, the server V and the server W can also communicate a common session key after they perform challenge-response authentication with each other successfully. Not only can this session key be used for later communication, but it can also provide for symmetric encryption/decryption usage. In the appendix, we provide a formal security proof of the session key.

5.6. Forward Secrecy Proof

In the EHR migration phase, a patient can delegate the server V to authenticate with the desired server of hospital W. They can then build the session key after successful authentication. In fact, they can use this session key to communicate with each other to transfer the patient U's EHRs or update the patient U's EHR. With this property, the system can reduce the communication bits and improve the efficiency of data transmission. In the appendix, we also provide a formal secrecy proof of the session key.

5.7. EHR Fair Exchange

In the EHR migration phase, if W does not receive the U's EHR from the V or if U's EHR is broken, then the patient U can perform the data recovery request to the HCA and ask the HCA for help to solve this situation by providing the above signatures and V's receipt to HCA. If the above signatures are valid, HCA performs the data recovery phase and forwards the encrypted patient's EHR to the system of the hospital W. Finally, the server of hospital W can also obtain the patient U's EHR under the help of HCA.

5.8. Offline Trusted Third Party

In the proposed scheme, we assume that there is a HCA and that it generates the patient's EHR migrating signature with a delegation document and performs data recovery. Here we can assume that the on-line device of the HCA can generate the signature after verifying the request party's signature in the migration registration phase. Only if there is a request coming in the data recovery phase would the HCA be on-line and solve this situation after verifying the request party's evidence, including the registration signatures and the related signatures. From the above setting, our trusted third party would not stay on-line all the time and just appears when it is needed. Additionally, only the HCA knows the link information (UID_U, ID_U) of the patient U. Therefore, the patient can prevent their real identity from being disclosed during the EHR migration transaction.

From the above security analysis properties, we take [10] as a reference and make comparisons with schemes from [6–10]. In the following, we provide some security analysis definitions for security comparison (Table 1).

Table 1. Security comparison.

Attributes	[6]	[7]	[8]	[9]	[10]	Ours
A1	Y	Y	Y	Y	Y	Y
A2	N	N	Y	Y	Y	Y
A3	Y	Y	Y	Y	Y	Y
A4	N	N	N	N	N	Y
A5	Y	Y	Y	Y	N	Y
A6	Y	N	N	N	N	Y
A7	N	N	N	N	N	Y
A8	N	N	N	N	N	Y

A1: Replay Attack Resistance; A2: Resist User Impersonation Attack; A3: Provide Mutual Authentication, A4: Provide Data Security; A5: Session Key Establishment, A6: Forward Secrecy Proof; A7: EHR Fair Exchange, A8: Offline Trusted Third Party.

5.9. Efficiency Comparisons

In this section, we evaluate our proposed scheme's efficiency. First, we assume that our scheme's parameter p is of 1024 bits for security consideration. We assume that H is the computation time of one hashing operation, Exp is the computation time of one modular exponential operation in a 1024 bit module, M is the computation time of one modular multiplication in a 1024 bit module, EC_M is the computation time of a number over an elliptic curve, and EC_P is the computation time of a bilinear pairing operation of two elements over an elliptic curve in [13–15]. We also let Sig, ASE, ADE, SE, and SD be the signature operation time, the asymmetric encryption time, the asymmetric decryption time, the symmetric encryption time, and the symmetric decryption time, respectively. We assume that our proposed scheme can be implemented on an elliptic curve over a 163-bit field and has the same security level of a 1024 bit public key crypto-system such as RSA or the Diffile-Hellman cryptosystem. We also assume that $Exp = 8.24EC_M$ for the ARM CPU to the processor in 200 Mhz [15]. We also determine certain relations from the following: $Exp \approx 240M = 600H \approx 3EC_P$, and $EC_A \approx 5M$ in [16–22].

Based on [23], a public key encryption/decryption operation time in an elliptic curve is approximately $1EC_A$ and $1EC_M + 1EC_A$, respectively. Therefore, our proposed scheme total computation time cost is about $9H + 3Sig + 14\oplus + 2ASE + 1ADE \approx 60.075M + 14\oplus$. Due to the different properties of the above schemes, we omitted the efficiency comparisons and found some currently survey papers [11,24] that have the same functional properties as our proposed scheme.

In [11], the authors proposed a dynamic consent model of health data sharing using blockchain technology. They combine the consent representation models (DUO) and ADA-M [24] to let patients control their EHR sharing to match the request query with full access rights. Their method is designed for building an EHR platform but is not a practical mechanism for patients exchanging their EHRs with a formal security proof in a blockchain environment. In [12], the authors proposed an EHR with a patient-centric access right framework model by using blockchain technology. We think that this is a good idea for building health information exchange systematic modules with blockchain in the future, but they do not offer a practical solution for EHR migration currently, even in a blockchain environment. Our proposed scheme is established by the functional block such as the signature functions with other authentication functions. In future work, our proposed scheme could functionally add a smart contract function to generate a verifiable functional patient EHR block in blockchain network. Hence, our proposed scheme could be used in blockchain and non-blockchain environments.

In the efficiency evaluation of our scheme, we used a desktop with Ubuntu 20.04 with Intel(R) Core(TM) i7-8700 CPU @ 3.20 GHz CPU and 15 GB memory. The simulation experiment was carried out using GO language, and the standard "crypto/elliptic" library was used. We simulated every phase 20 times, shown in Figures 4–6.

In the future, we will discuss the forged HCA problem [25] and other applications such as neural network environments for COVID-19 patients [26] exchanging their EHRs. We hope to have a good solution to the above problems.

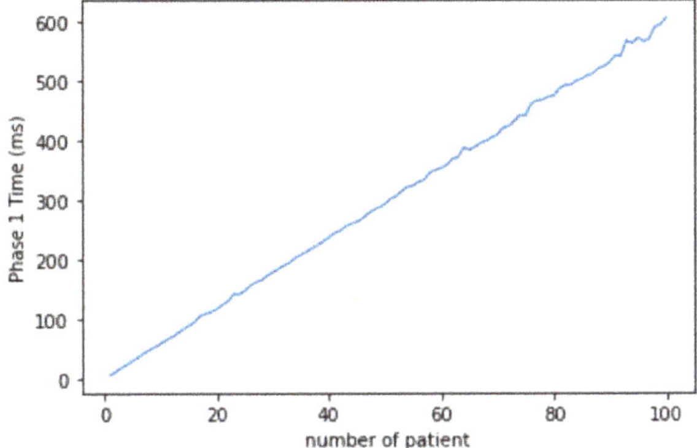

Figure 4. The migration registration phase simulation.

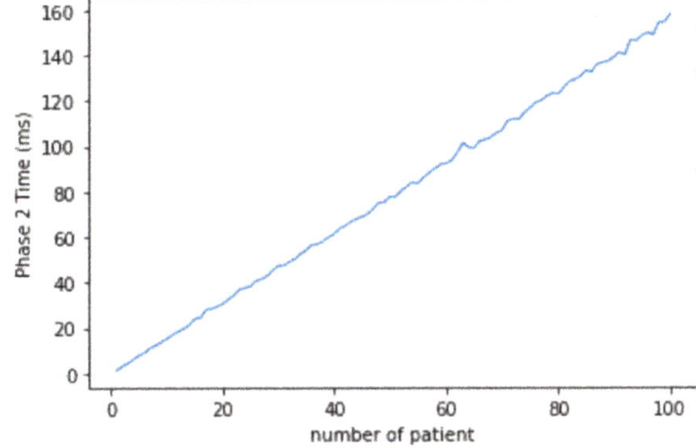

Figure 5. The EHR migration phase simulation.

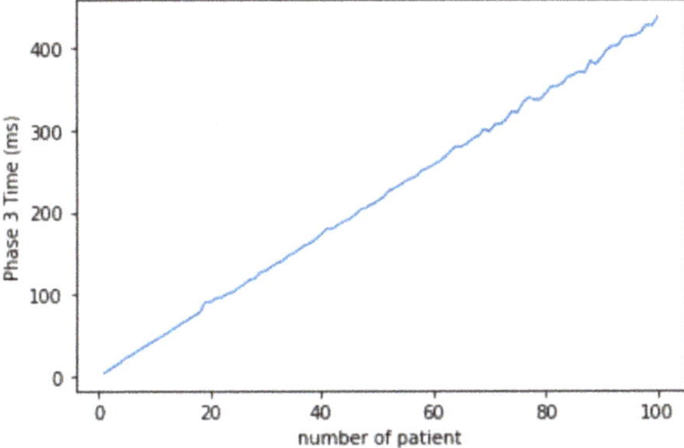

Figure 6. The data recovery phase simulation.

6. Security Proof

In this section, we continue to demonstrate what an adversary is and its probability. We model the *Game*, our scheme simulation steps, and the related oracle responses.

An adversary (call it $\mathcal{S}$) can control all communication messages in this scheme. The adversary can obtain related information by sending oracles. A *Game* is the simulation of our proposed scheme, which can equip all kinds of oracles, and oracles can reply back according to the adversary's questions. There is also another adversary (call it $\mathcal{S}$) that controls the simulation and takes $\mathcal{A}$'s ability to break the hard problem defined in the security definition.

Let *Game* be a "game", the simulation of our scheme, where the adversary $\mathcal{A}$ can ask queries to the oracles, and the oracles can answer back to the adversary. The following are query types that an adversary can make in the game.

- Send query $Send(i,k,M)$(or $Send(j,k',M)$): this query models an adversary that can send message M to the authentication party i(or j), where $i,j \in \{U,V\}$ in the k- (or k')-th session, where k and k' are two different session numbers in N.
- Reveal query $Reveal(i,k)$(or $Reveal(j,k)$): this query is used to model a situation that exposes a session key of Π_i^k(or $\Pi_j^{k'}$) to an adversary, where $i,j \in \{U,V\}$ and $k,k' \in N$ in the k(or k')-the session.
- Encryption query: this query is used to model that an adversary can obtain a ciphertext C' on the input of a chosen message M'.
- Decryption query: this query is used for modelling an adversary that can obtain a plain-text M' on the input of a cipher-text C'.
- Corrupt query $Corrupt(i)$(or $Corrupt(j)$): this query is used to expose the private key of the player p_i(or p_j) to the adversary, where $i,j \in \{U,V\}$.
- Hash query: this query depends on what the input is; the simulator then returns the related output to the attacker.
- Test query $Test(i,k)$: this query is used to define the advantage of an adversary. When the adversary $\mathcal{A}$ has finished all of the above queries to the oracle, they can make this test query on an instance Π_i^k(or $\Pi_j^{k'}$) to the simulator. At this time, the simulator will flip a coin b. If b is 1, then the real session key is $ssk_{i,j}^k$. Otherwise, it returns a random string chosen uniformly from $\{0,1\}^*$. The adversary is only allowed to ask for the "fresh" instance of a player in the above simulation.

Proof of Theorem 1. First, we assume that there is an adversary $\mathcal{A}$ that attempts to attack our patient EHR exchange scheme (*PEHRES*) in the forward secure sense. We then let *dis* be the event at which $\mathcal{A}$ can distinguish at least one ciphertext in *PEHRES* with non-negligible probability. At the same time, we also let *forge* be the event at which the adversary $\mathcal{D}$ can forge the signature of our *PES* with non-negligible probability. We assume that

$$Pr_{\mathcal{A}}[b=b'] \leq Pr_{\mathcal{A}}[b=b' \wedge \overline{dis} \wedge \overline{forge}] + Pr_{\mathcal{A}}[dis] + Pr_{\mathcal{A}}[forge],$$

where b and b' are coin flips chosen by the simulator and the attacker $\mathcal{A}$, respectively.

We also assume that

$$Pr_{\mathcal{F}^*}[forge] \leq Pr_{\mathcal{F}^*}[\mathcal{F}^* \rightarrow S_U^* | Ver(S_U^*) = 1] + Pr_{\mathcal{F}^*}[\mathcal{F}^* \rightarrow S_{HCA}^* | Ver(S_{HCA}^*) = 1],$$

where S_U^* and S_{HCA}^* are signatures forged by the attacker $\mathcal{F}^*$, respectively. We then use three lemmas to complete this security proof in the following.

Lemma 4. *We assume that there is no event such that the attacker $\mathcal{A}$ can distinguish the ciphertext C^* with non-negligible probability*

$$Pr_{\mathcal{A}}[dis] \leq \frac{1}{2}(I^2 q_h q_e q_s (Adv_{ASE,\mathcal{D},C_{HCA}^*}^{Ind-CCA}(\theta,t')) + 1) + \frac{1}{2}(I^2 q_h q_e (Adv_{ASE,\mathcal{D},C_V^*}^{Ind-CCA}(\theta,t')) + 1),$$

in the polynomial time bound t' under the above Ind-CCA security definition with q_h hash queries, at most q_e encryption queries, and at most q_s decryption queries, respectively.

Proof of Lemma 4. We assume that $Pr[dis]$ is a non-negligible probability in the simulation game. We can then construct an attacker $\mathcal{D}$ whose work is to distinguish the cipher-text under the Ind-CCA encryption/decryption scheme. There is also an attacker $\mathcal{F}$ whose goal is to break the encryption/decryption of our proposed scheme SE. Next, we construct $\mathcal{D}$ as the simulator that simulates the attacking environment in which $\mathcal{F}$ can mount its attack. First, $\mathcal{D}$ simulates an encryption oracle $SE_{pk_i}(\cdot, \theta)$, where $i \in \{U, V\}$, and generates the C' to the attacker $\mathcal{F}$ on the plain-texts (M') chosen by the attacker $\mathcal{D}$ in the selected instance $\Pi^k_{i^*}$, where the partner of $\Pi^k_{i^*}$ is p_{j^*}. In addition, $\mathcal{D}$ also simulates the decryption oracle to answer the decryption query issued by the attacker $\mathcal{D}$. We consider the following steps. First, $\mathcal{D}$ prepares all hash functions, including H_1 and H_2, two hash functions with collision-resistance. It also generates the instances $i^*, j^* \longleftarrow [1, ..., I-1]$ of each player i, where $i \in \{U, V\}$. It can make the above two hash queries l^* times, where $l^* \longleftarrow [1, ..., q_h]$.

Hash query

In this hash query phase, the simulator also responds to all kinds of hash queries in each stage.

- In the migration registration phase, the simulator generates the corresponding hashed value to U and V. It prepares the $H_1(Bio_U)$ for the patient U as the registration token. At the same time, the simulator chooses r''_U and makes the ciphertext $C_{HCA} = (r''_U, Bio_U, UID_U, Cert_U, ID_U, EHR_U \oplus r''_U)$ for $\mathcal{F}$.
- Next, the simulator has to simulate the signature S_U and S_{HCA} from the signing oracle. It then forwards (S_U, S_{HCA}) to $\mathcal{F}$. The simulator then computes $Sig_V(S_{HCA}, S_U, Agree_{V->W})$ as the response receipt of the instance of player V.
- In the EHR migration phase, the simulator can simulate the hashed values $H_2(r^*_V + 1)$ and $H_2((r^*_W + 1) \oplus r''_U)$ to $\mathcal{F}$, which forwards $r^*_V + 1$, $(r^*_W + 1)$, and r''_U as challenge random numbers.

Phase 1

- In the migration registration phase, the attacker $\mathcal{F}$ can issue the encryption query on the chosen message $M' = (r''_U, Bio_U, UID_U, Cert_U, ID_U, EHR_U \oplus r''_U)$. $\mathcal{D}$ can then forward this M' to the encryption oracle and pass the final result C' to the $\mathcal{F}$.
- The attacker $\mathcal{F}$ can issue the decryption query on the ciphertext C'. $\mathcal{D}$ can then forward this C' to the decryption oracle and pass the final message $M' = (r''_U, Bio_U, UID_U, Cert_U, ID_U, EHR_U \oplus r''_U)$ from the oracle output to the $\mathcal{F}$.
- In the EHR migration phase, the attacker $\mathcal{F}$ can ask for the $M'' = r^*_V$ encryption result C''_V and the decryption result of M''. $\mathcal{D}$ can forward C''_V and M'' to the attacker $\mathcal{F}$.

Challenge

- In this phase, $\mathcal{D}$ can generate the ciphertext C^*_{HCA} by querying the asymmetric encryption oracle $ASE_{pk_T}(M^*_b, \theta)$ with the coin flip b on the target message pair (M^*_0, M^*_1) chosen by the attacker $\mathcal{F}$. If b=1, C is computed from $ASE_{pk_T}(M^*_1, \theta)$, where $T \in \{U^*, V^*\}$. Otherwise, it returns the ciphertext C^* to $\mathcal{F}$, where $C^*_{HCA} \longleftarrow ASE_{pk_T}(M^*_0, \theta)$. The only restriction is that $\mathcal{F}$ cannot ask for the decryption query on the ciphertext C^*_{HCA}. On the other hand, we also consider that the ciphertext C^*_V is the same situation. We also set up the target message pair (M^{**}_0, M^{**}_1) chosen by the attacker $\mathcal{F}$. If b'=1, C is computed from $ASE_{pk_T}(M^{**}_1, \theta)$, where $T \in \{W^*\}$. Otherwise, it returns the ciphertext C^*_V to $\mathcal{F}$, where $C^*_V \longleftarrow ASE_{pk_T}(M^{**}_0, \theta)$.
- We assume that this event dis happens with respect to the instance $\Pi^{l^*}_{i^*}$ of the player i, where its partner player is p_{j^*}. At this time, $\mathcal{F}$ finally outputs its own guessing bit b'. Otherwise, the system stops this authentication stage and aborts this simulation.

Finally, $\mathcal{F}$ has a set with instances of players i^* and j^* with q_h total hash queries , at most q_e encryption queries, and q_s decryption queries. At this time, $\mathcal{D}$ does not fail in the simulation environment with $\mathcal{F}$'s correct guessing, where $b = b'$ has non-negligible probability. The following equation will then hold:

$$\begin{aligned} Adv^{Ind-CCA}_{ASE,\mathcal{D},C^*_{HCA}}(\theta,t') \leq \\ \frac{1}{I^2 q_h q_e q_s}(Pr[Game^{Ind-CCA-1}_{ASE,\mathcal{F}}(\theta)=1] - Pr[Game^{Ind-CCA-0}_{ASE,\mathcal{F}}(\theta)=1]) = \\ \frac{1}{I^2 q_h q_e q_s}(Pr[Game^{Ind-CCA-1}_{ASE,\mathcal{F}}(\theta)=1] - (1 - Pr[Game^{Ind-CCA-1}_{ASE,\mathcal{F}}(\theta)=1])) = \\ \frac{1}{I^2 q_h q_e q_s}(2(Pr[Game^{Ind-CCA-1}_{ASE,\mathcal{F}}(\theta)=1]) - 1). \end{aligned} \quad (5)$$

In the ciphertext C^*_V simulation game, we have the same simulation as above. Therefore, we omitted the simulation, but we also conclude that

$$Adv^{Ind-CCA}_{ASE,\mathcal{D},C^*_V}(\theta,t') \leq \frac{1}{I^2 q_h q_e}(2(Pr[Game^{Ind-CCA-1}_{ASE,\mathcal{F}}(\theta)=1]) - 1) \quad (6)$$

We then can summarize the total probability as follows:

$$\begin{aligned} Pr_{\mathcal{A}}[dis] \leq Pr[Game^{Ind-CCA-1}_{ASE,\mathcal{F}}(\theta)=1] \leq \\ \frac{1}{2}(I^2 q_h q_e q_s (Adv^{Ind-CCA}_{ASE,\mathcal{D},C^*_{HCA}}(\theta,t')) + 1) + \frac{1}{2}(I^2 q_h q_e (Adv^{Ind-CCA}_{ASE,\mathcal{D},C^*_V}(\theta,t')) + 1). \end{aligned} \quad (7)$$

□

Lemma 5. *Before we prove this lemma, we assume that there is no attacker $\mathcal{A}$ that can guess the real session key in the event that the ciphertext C^* generated by the symmetric encryption (SE) functions cannot be distinguished by $\mathcal{A}$ correctly with non-negligible probability. We then have*

$$Pr_{\mathcal{A}}[b = b' \wedge \overline{dis} \wedge \overline{forge}] \leq \frac{1}{2}((Iq_h)^2 Adv^{Ind}_{\mathcal{A},SE}(\theta,t^*) + 1),$$

in the polynomial time t^* under the random oracle (RO) assumption with total q_h hash queries.

Proof of Lemma 5. In this proof, we construct another simulator $\mathcal{C}$ that also simulates the attacking environment for $\mathcal{A}$ mounting its attack. Finally, if $\mathcal{A}$ can guess the real session key successfully with the non-negligible property, then we can use $\mathcal{A}$ to break the random oracle assumption.

- First, we assume that $\mathcal{C}$ is given the system parameters $(G, g, q, H_1, H_2, t_{i^*}, t_{j^*})$. It starts to choose public key/secret key pairs for all parties except for p_{i^*} and p_{j^*}. Next, $\mathcal{C}$ selects other protocol parameters such as $(i^*, j^*) \longleftarrow [1, ..., I-1]$ and $t_1, t_2 \longleftarrow [1, ..., q_h]$. At this time, we also let the target identities i^* and j^* be the instances of the patient U and the system V, respectively.
- After the above environment is set up completely, $\mathcal{C}$ starts to simulate the following oracle queries and related hash functions.
- First, $\mathcal{C}$ sets parameters $(i, j, r_i, r_j, AE_{pk_T}(M_b, \theta))$, where r_i, r_j are two random numbers, and H_1,H_2 are these two collision-resistance hash functions. In addition, $\mathcal{C}$ adopts θ as the security parameter. First, $\mathcal{C}$ prepares two nonce challenge numbers r_i and r_j in the t_1-th and t_2-th session, respectively.
- During this simulation, for each query issued from $\mathcal{A}$, $\mathcal{C}$ answers it as follows:
- It takes $(i, j, H_1, H_2, \omega_1, \omega_2, r_i, r_j)$ as its input and responds to each *Send* query in the instance Π^k_i in the k-th session on the message M, where ω_1 and ω_2 are two pseudo-random functions with the same length of the hash oracles H_1 and H_2, respectively.

□

Hash Query

In this hash query phase, the simulator can answer all kinds of harsh queries in each stage, as follows:

- In the migration registration phase, the simulator also computes the initial biometric information value $(r_{B_U} \oplus H_1(Bio_U))$ for the patient U and keeps them in the oracle simulation database. If the $\mathcal{A}$ makes the hash query on $(r_V + 1||r_{HCA})$ and $r_{HCA} + 1$, $\mathcal{C}$ also forwards them to the hash oracle and returns the response hashed value to $\mathcal{A}$.
- In the EHR migration phase, $\mathcal{A}$ asks for the $r_V^* + 1$, $(r_W^* + 1) \oplus r_U''$, and $(r_W^* + 1)||(r_V^* + 1)$ hash values of the H_2 function, and $\mathcal{C}$ also forwards them to the hash oracle and lets the hash oracle generate the corresponding hash values to $\mathcal{A}$.
- If $\mathcal{C}$ receives the $Corrupt(i)$ query, then it returns the private key to $\mathcal{A}$. If $\mathcal{C}$ receives the $Reveal(i)$ query, it checks if $i \neq i^*$ or $j \neq j^*$, then $\mathcal{D}$ computes the session key $ssk_{i,j} = H_1(\omega_1(r_W^* + 1))||\omega_2((r_V^* + 1)))$, where $r_W^* + 1$ and $r_V^* + 1$ are chosen from ω_1 and ω_2 functions, respectively. On the other hand, if $i = i^*$, $j = j^*$, and $t_1 = t_2 = l$, then $\mathcal{C}$ computes $H_1((r_W^* + 1)||(r_V^* + 1))$ as the session key $ssk_{i,j}$ and delays this key in the response in the $Test$ query.
- If the adversary $\mathcal{A}$ has finished the above queries, it can make the $Test$ query to $\mathcal{C}$. At this time, $\mathcal{C}$ will check the instance session and player to see if $i = i^*$ and $j = j^*$, and $\mathcal{C}$ then tosses a coin b to answer the session key. If b=0, then it computes $ssk_{i,j} = H_1((r_W^* + 1)||(r_V^* + 1))$. Otherwise, it computes $ssk_{i,j} \longleftarrow \{0,1\}^*$ from the random pseudo-random function.

Finally, if $\mathcal{C}$ answers the $Test$ query for $\Pi_{i^*}^{t_1}$ and $\Pi_{j^*}^{t_2}$ by using $(Z_n^*, H_1, H_2, \omega_1, \omega_2)$, and $\mathcal{A}$ does not fail in guessing b', then $\mathcal{A}$ answers the session key depending on its coin flip b'. We can have

$$\begin{aligned} &Adv_{\mathcal{C},\mathcal{A},SE}^{H_1,H_2,\omega_1,\omega_2}(\theta, t) = \\ &Pr[\mathcal{C}(Z_n^*, H_1, H_2, \omega_1, \omega_2) = 1|ssk_{i,j} = H_1((r_W^* + 1)||(r_V^* + 1))))]- \\ &Pr[\mathcal{C}(Z_n^*, H_1, H_2, \omega_1, \omega_2) = 1|ssk_{i,j} \longleftarrow \{0,1\}^*, t \in Z_q^*] \leq \\ &\frac{1}{(Iq_h)^2}(Pr[\mathcal{A}(\cdot) = 1|ssk_{i^*,j^*} \text{ is real in } Test \text{ query}] - Pr[\mathcal{A}(\cdot) = 1|ssk_{i^*,j^*} \text{ is random in } Test \text{ query}]) \leq \\ &\frac{1}{(Iq_h)^2}(2Pr_{\mathcal{A}}[b = b' \wedge \overline{dis} \wedge \overline{forge}] - 1). \end{aligned} \tag{8}$$

Finally, we could conclude that

$$Pr_{\mathcal{A}}[b = b' \wedge \overline{dis} \wedge \overline{forge}] \leq \frac{1}{2}((Iq_h)^2 Adv_{\mathcal{C},\mathcal{A},SE}^{H_1,H_2,\omega_1,\omega_2}(\theta, t) + 1). \tag{9}$$

Lemma 6. *Before we prove Lemma 1, we assume that there is no event such that the attacker $\mathcal{F}^*$ can forge the signature S_U of patient U with non-negligible probability*

$$Pr_{\mathcal{F}}[forge] \leq (l^3 q_s(Adv_{Sig,\mathcal{S},\mathcal{F}}^{Unf}(\theta, t^*),$$

in the polynomial time bound t^* under the above Ind-CCA security definition with q_h hash queries, at most q_e encryption queries, and at most q_s decryption queries, respectively.

Proof of Lemma 6. In this lemma proof, we start to prove our above **Lemma 1** (Unforgeability). To start the proof of **Lemma 1** (Unforgeability), we defined the **Game** as the simulation game that runs as the proposed protocol controlled by the simulator $\mathcal{S}$. We define **Game** as follows.

$\mathbf{Game}^{Unf}_{\mathcal{A},Sig}(\theta, t)$
Phase 1.
$\mathcal{F} \longleftarrow \{M\}^l$
$S_i \longleftarrow \mathcal{F}^{Sig(sk_i,M_i),RO_1,RO_2,H_1,H_2}(\theta, t)$
Challenge Phase.
$i \in \{U, HCA\}, M^* \longleftarrow \mathcal{F}(M)$,
Loop j=1 to l
$S'_{i,j} \longleftarrow \mathcal{F}^{Sig(sk_i)}(\theta, t)(M^*)$
If (Ver($S'_{i,j+1}$) == 1 and $S'_{i,j+1} \notin S_{i,j}$).
Break
Return $S'_{i,j+1}$.
else if (j <= l)
goto Loop

We first define the simulator $\mathcal{S}$ as the simulator that is given in the RSA factoring problem, and we assume there is an attacker $\mathcal{F}$ whose goal is to forge a valid signature on the Sig function block.

The simulator $\mathcal{S}$ first chooses the security parameter l with the message space M^l. The $\mathcal{S}$ also selects two collision-resistance hash functions that map from $Z_n^* \longrightarrow \{0,1\}$ and two hash oracles RO_1 and RO_2, respectively. After setting up the system parameter, the $\mathcal{S}$ simulates each phase in the proposed EHR scheme. In the migration registration phase, the attacker $\mathcal{F}$ can impersonate the patient U to ask for U's signature request S_U on the desired message. When $\mathcal{S}$ has received this request, it takes the message as input and outputs the signature S_U with the help of the above secure digital signature function Sig. It then returns this S_U back to the $\mathcal{F}$. The $\mathcal{F}$ can also continue to ask the hospital certification center HCA's signature on the received signature S_{HCA} of patient P. It also receives the message tuple $(S_U, Date, ID_U)$ and outputs the signature S_{HCA} back to $\mathcal{F}$. In addition, it is the same situation when $\mathcal{F}$ asks the signature of V. The $\mathcal{S}$ also returns S_V back to $\mathcal{F}$.

In these phases, the $\mathcal{F}$ will make the signature request in the above situation. The $\mathcal{S}$ starts the **Challenge phase** and forwards the message M_i, where $i \in \{U, V, HCA\}$. The $\mathcal{F}$ can forge l+1 signatures $S'_{i,j}$, where $S'_{i,j+1} \notin S_{i,j} \longleftarrow Sig(sk_i, M^*)$, where $i \in \{U, V, HCA\}$ and $j = 1 \sim l$ after l signature queries. Finally, this forged signature also passes the verification $Ver(S'_{i,j+1})$ successfully. We then can use $\mathcal{F}$'s ability to find a solution to the RSA factoring problem. Thus, we have

$$\begin{aligned} Adv^{Unf}_{Sig,\mathcal{S},\mathcal{F}}(\theta, t^*) \geq & |Pr[S'_{i,j+1} \longleftarrow \mathcal{F}^{Unf}_{Sig}(M^*), Ver(S'_{i,j+1} = 1)]| \\ = & \frac{1}{I^3 q_s}(Pr[S'_{i,j+1} \longleftarrow \mathcal{F}^{Unf}_{Sig}(M^*), Ver(S'_{i,j+1} = 1)]). \end{aligned} \tag{10}$$

Finally, we could conclude that

$$Pr_{\mathcal{F}}[forge] \leq (I^3 q_s) Adv^{Unf}_{Sig,\mathcal{S},\mathcal{F}}(\theta, t^*).$$

□

After summarizing the above three lemmas, we can conclude that $\frac{1}{2}(I^2 q_h q_e q_s (Adv^{Ind-CCA}_{SE,\mathcal{D},C^*_{HCA}}(\theta, t')) + 1) + \frac{1}{2}(I^2 q_h q_e (Adv^{Ind-CCA}_{SE,\mathcal{D},C^*_V}(\theta, t')) + 1) + \frac{1}{2}((I q_h)^2 Adv^{Ind}_{\mathcal{A},SE}(\theta, t^*) + 1) + (I^3 q_s) Adv^{Unf}_{Sig,\mathcal{S},\mathcal{F}}(\theta, t^*)$. □

7. Conclusions

We propose a practical and provable patient EHR fair exchange scheme with key agreement for e-health information systems. Not only does our scheme offer a solution for the seven problems described in Section 2, when a patient attempts to migrate their personal information data to another hospital, but they can also maintain their anonymity during the data migration transaction. In addition, Table 1 shows a security and functional comparison with other related papers. It is obvious that our proposed scheme guarantees convenience, rapidity, and integrity.

Our mechanism provides secure data storage and the secure transfer of authorized information to designated locations. What information can be authorized, for example, whether COVID-19 patient privacy concerning patients' names, identities, and genetic sequences can be transmitted to different hospitals, is beyond this study's scope. This study guarantees secure data transfer and storage. Our scheme also provides a formal security proof in the random oracle model under chosen-ciphertext security. Our approach focuses on the security and privacy protection of patient EHRs rather than on the design of electronic health systems. It not only serves as a high-level functional module for integrity but also provides an efficient and contactless data transfer method that allows for medical data aggregation and protects patient anonymity, especially relevant in the context of the global COVID-19 pandemic. In the future, we will extend our scheme to be applicable for COVID-19 patient EHR exchange in a neural network environment.

Author Contributions: Conceptualization, M.-T.C.; Formal analysis, M.-T.C. and T.-H.L.; Methodology, M.-T.C. and T.-H.L.; Writing—original draft, M.-T.C.; Writing—review & editing, T.-H.L. All authors have read and agreed to the published version of the manuscript.

Funding: This research received no external funding.

Institutional Review Board Statement: Not applicable.

Informed Consent Statement: Not applicable.

Acknowledgments: This study was supported in part by grants from the Ministry of Science and Technology of the Republic of China (Grant No. MOST 109-2221-E-167-028-MY2).

Conflicts of Interest: The authors declare no conflict of interest.

References

1. Saranummi, N. In the spotlight: health information systems. *PHR Value Based Healthc.* **2011**, *2*, 15–17. [CrossRef] [PubMed]
2. Chiuchisan, I.; Balan, D.G.; Geman, O.; Chiuchisan, I.; Gordin, I. A Security Approach for Health Care Information Systems. In Proceedings of the E-Health and Bioengineering Conference (EHB), Sinaia, Romania, 22–24 June 2017; pp. 721–724.
3. Appari, A.; Johnson, M. Information security and privacy in healthcare: Current state of research. *Int. J. Internet Enterp. Manag.* **2010**, *4*, 279–284. [CrossRef]
4. Mahmoud, A.M.; Zeki, A.M. Security issues with health care information technology. *Int. J. Sci. Res. (IJSR)* **2015**, *12*, 1021–1024.
5. Health Level Seven. Available online: http://www.hl7.org/implement/standards/ansiapproved.cfm (accessed on 21 December 2020).
6. Das, A.K.; Goswami, A. A secure and efficient uniqueness- and-anonymity-preserving remote user authentication scheme for connected health care. *J. Med. Syst.* **2013**, *3*, 9948. [CrossRef] [PubMed]
7. He, D.; Kumar, N.; Chen, J.; Lee, C.C.; Chilamkurti, N.; Yeo, S.S. Robust anonymous authentication protocol for health- care applications using wireless medical sensor networks. *Multimed. Syst.* **2015**, *1*, 49–60. [CrossRef]
8. Amin, R.; Islam, S.K.H.; Biswas, G.P.; Khan, M.K.; Kumar, N. A robust and anonymous patient monitoring system using wireless medical sensor networks. *Future Gener. Comput. Syst.* **2018**, *80*, 483–495. [CrossRef]
9. Zhang, L.; Zhang, Y.; Tang, S.; Luo, H. Privacy protection for e-health systems by means of dynamic authentication and three-factor key agreement. *IEEE Trans. Ind. Electron.* **2017**, *65*, 2795–2805. [CrossRef]
10. Kaul, S.D.; Murty, V.K.; Hatzinakos, D. Secure and privacy preserving biometric based user authentication with data access control system in the healthcare environment. In Proceedings of the 2020 International Conference on Cyberworlds (CW), Caen, France, 29 Septemeber–1 October 2020; pp. 249–256.
11. Jaiman, V.; Urovi, V. A Consent Model for Blockchain-Based Health Data Sharing Platforms. *IEEE Access* **2020**, *8*, 143734–143745. [CrossRef]

12. Zhuang, Y.; Sheets, L.R.; Chen, Y.W.; Shae, Z.Y.; Tsai, J.J.P.; Shyu, C.R. A Patient-Centric Health Information Exchange Framework Using Blockchain Technology. *IEEE J. Biomed. Health Inform.* **2020**, *24*, 2169–2176. [CrossRef] [PubMed]
13. Jurisic, A.; Menezes, A.J. Elliptic Curves and Cryptography; pp. 1–13. Available online: http://http://www.cs.nthu.edu.tw/~cchen/CS4351/jurisic.pdf (accessed on 22 December 2020).
14. Koblitz, N.; Menezes, A.; Vanstone, S. The state of Elliptic curve cryptography. *Des. Codes Cryptgography* **2000**, *19*, 173–193. [CrossRef]
15. Lauter, K. the Advantages of Elliptic curve cryptography for wireless security. *IEEE Wirel. Commun.* **2004**, *11*, 62–67. [CrossRef]
16. LI, Z., Higgins, J., Clement, M., Performance of Finite Field Arithmetic in an Elliptic Curve Cryptosystem. In Proceedings of the 9th IEEE International Symposium on Modeling, Analysis, and Simulation of Computer and Telecommunications Systems (MASCOTS'01), Cincinnati, OH, USA, 15–18 August 2001; pp. 249–256.
17. Ramachanfdran, A.; Zhou, Z.; Huang, D. Computing cryptography algorithm in Portable and embedded devices. In Proceedings of the IEEE International Conference on Portable Information Devices, Orlando, FL, USA, 25–29 May 2007; pp. 1–7.
18. Schneier, B. *Applied Cryptography*, 2nd ed.; John Wiley & Sons: Hoboken, NJ, USA, 1996.
19. Takashima, K. Scaling Security of Elliptic Curves with Fast Pairing Using Efficient Endomorphisms. *IEICE Trans. Fundam.* **2007**, *E90-A*, 152–159. [CrossRef]
20. Bertinoi, G.; Breveglieri, L.; Chen, L.; Fragneto, P.; Harrison, K.; Pelosi, G. A pairing SW implementation for smart cards. *J. Syst. Softw.* **2008**, *81*, 1240–1247. [CrossRef]
21. Hankerson, D.; Menezes, A.; Scott, M. Software Implementation of pairings. *Identity-Based Cryptogr. Cryptol. Inf. Secur.* **2008**, *2*. [CrossRef]
22. Hohenberger, S. Advances in Signatures, Encryption, and E-cash from Bilnear Groups. Ph.D. Thesis, Massachusetts Institute of Technology, Cambridge, MA, USA, 2006.
23. Juang, W.S.; Lei, C.L.; Liaw, H.T.; Nien,W.K. Robust and Efficient Three-party User Authentication and key agreement Using Bilinear pairings. *Int. J. Innov. Comput. Inf. Control.* **2010**, *6*, 763–772.
24. Woolley, J.P.; Kirby, E.; Leslie, J.; Jeanson, F.; Cabili, M.N.; Rushton, G.; Hazard, J.G.; Ladas, V.; Veal, C.D.; Gibson, S.J.; et al. Responsible sharing of biomedical data and biospecimens via the automatable discovery and access Matrix (ADA-M). *NPJ Genom. Med.* **2018**, *3*, 7. [CrossRef] [PubMed]
25. Chakraborty, T.; Jajodia, S.; Katz, J.; Picariello, A.; Sperli, G.; Subrahmanian, V.S. FORGE: A fake online repository generation engine for cyber deception. *IEEE Trans. Dependable Secur. Comput.* **2019**. [CrossRef]
26. La Gatta, V.; Moscato, V.; Postiglione, M.; Sperli, G. An epidemiological neural network exploiting dynamic graph structured data applied to the COVID-19 outbreak. *IEEE Trans. Big Data.* **2020**, *7*, 45–55. [CrossRef]

MDPI

Article

A Data Driven Approach for Raw Material Terminology

Olivera Kitanović [1,*,†], Ranka Stanković [1,†], Aleksandra Tomašević [1,†], Mihailo Škorić [1,†], Ivan Babić [2,†] and Ljiljana Kolonja [1,†]

1 Faculty of Mining and Geology, University of Belgrade, 11000 Belgrade, Serbia; ranka.stankovic@rgf.bg.ac.rs (R.S.); aleksandra.tomasevic@rgf.bg.ac.rs (A.T.); mihailo.skoric@rgf.bg.ac.rs (M.Š.); ljiljana.kolonja@rgf.bg.ac.rs (L.K.)
2 Department for Informatics and Computing, University of Criminal Investigation and Police Studies, 11000 Belgrade, Serbia; ivan.babic@mup.gov.rs
* Correspondence: olivera.kitanovic@rgf.bg.ac.rs; Tel.: +381-11-3219-212
† These authors contributed equally to this work.

Abstract: The research presented in this paper aims at creating a bilingual (sr-en), easily searchable, hypertext, born-digital, corpus-based terminological database of raw material terminology for dictionary production. The approach is based on linking dictionaries related to the raw material domain, both digitally born and printed, into a lexicon structure, aligning terminology from different dictionaries as much as possible. This paper presents the main features of this approach, data used for compilation of the terminological database, the procedure by which it has been generated and a mobile application for its use. Available (terminological) resources will be presented—paper dictionaries and digital resources related to the raw material domain, as well as general lexica morphological dictionaries. Resource preparation started with dictionary (retro)digitisation and corpora enlargement, followed by adding new Serbian terms to general lexica dictionaries, as well as adding bilingual terms. Dictionary development is relying on corpus analysis, details of which are also presented. Usage examples, collocations and concordances play an important role in raw material terminology, and have also been included in this research. Some important related issues discussed are collocation extraction methods, the use of domain labels, lexical and semantic relations, definitions and subentries.

Keywords: raw material; mining; terminology; dictionary; terminology application; mobile application; digitization; lexical data; corpus data; linguistic linked open data

Citation: Kitanović, O; Stanković, R.; Tomašević, A.; Škorić, M.; Babić, I.; Kolonja, L. A Data Driven Approach for Raw Material Terminology. *Appl. Sci.* **2021**, *11*, 2892. https://doi.org/10.3390/app11072892

Academic Editor: Chuan-Ming Liu

Received: 25 February 2021
Accepted: 16 March 2021
Published: 24 March 2021

Publisher's Note: MDPI stays neutral with regard to jurisdictional claims in published maps and institutional affiliations.

Copyright: © 2021 by the authors. Licensee MDPI, Basel, Switzerland. This article is an open access article distributed under the terms and conditions of the Creative Commons Attribution (CC BY) license (https://creativecommons.org/licenses/by/4.0/).

1. Introduction

During the last decade, lexicography entered a new era due both to rapid development of advanced computational methods and availability of previously unseen abundance of language data in different modalities. These developments have opened new opportunities for producing modern Serbian monolingual and bilingual dictionaries, which will overcome the shortcoming of existing ones, characterized by obsolescence of macrostructure, microstructure and data presentation, frequent inaccuracy of translation, visual and typographic monotony, and a neglect of needs of potential users [1]. These new, modern dictionaries will enable potential users, including students, translators, teachers, researchers and other interested parties, to find all information on formal and contextual properties of words and their interrelationships, in one place. In addition to new human readable monolingual and bilingual dictionaries, machine readable dictionaries of both kinds are also needed. In this situation, a comprehensive approach, combining all available resources, which can be used for producing various types of dictionaries, especially in specialized and terminological domains, seem to be the optimal solution.

According to the findings of the Elexis project [2], the main positive changes in lexicography in the last 10–15 years are mostly related to digitisation and automation of lexicographic work, online publishing (moving from paper to online) and, with the beginning of the corpus era, by access to corpora supported by (semi)automatic extraction of terms. Automatic data extraction comprises data that is automatically obtained from corpora of authentic language use, which is then subjected to lexicographers' post-processing or included, as is, in the published dictionary, but marked as automatically derived from corpus data. It should be noted that data derived from existing lexical databases and dictionaries should be considered as reuse of data. One of the issues related is the processing and representation of terminological phrases, or multiword expressions (MWEs), ranging from compound nouns (e.g., nickname) to complex phrasal verbs (e.g., give up) and idiomatic expressions (e.g., break the ice), which has remained a challenge over the past 20+ years [3]. In our research we focused on semantically transparent terminological phrases, as well as terminological phrases that result in a meaning shift. Some frequent syntactic patterns and translation options will be discussed. In our approach we will use a combination of: reuse of data, automatic extraction and manual postediting.

The advantage of using online platforms, which offer the possibility of regular updates and a more effective collaboration via the internet, as well as the use of mobile devices were highlighted in literature [4]. The impact of mobile devices as a distribution method is immense, and a mobile-first approach is now instrumental. The general shift towards (mobile) life online brought a clear realization that "printed lexicography"—in general terms—is a thing of the past, and this also turned its business side upside down [5].

Wide adoption of mobile devices has created new ways of learning through interaction and communication and they are becoming integrated in the lives of today's students, enhancing mobility of the learning process. Thus, for example, Language for Specific Purposes (LSP) dictionaries are now being produced at the university level using mobile LSP lexicography. One such dictionary called MobiLex was produced at the Stellenbosch University in South Africa to enhance teaching and learning of historical terms, with favorable pedagogical consequences regarding the learning of such terms. Trends and developments in technology offer the possibility of changing the face of lexicographical support in a mobile environment, from a pedagogical perspective [6].

Big data analysis methods have opened new possibilities for analyzing corpora, which contain large amounts of textual data. Thus, for example, Chen et al. [7] propose a novel statistic-based corpus machine processing approach to refine big textual data, to be used for ESP (English for Specific Purposes). The approach is based on establishing a function word list and embedding it into the program, in order to refine the word list and keyword list. The aim is to enhance the efficiency of corpora processing, starting from preparatory work, followed by generating raw data, optimizing the process, and ending by generating refined data. COVID-19 news reports are used as a simulation example of big textual data and applied to verify the efficacy of the machine optimizing process.

Electronic lexicography offers important possibilities in comparison to the traditional approach. Examples of usage may be extracted from original texts and linked to dictionary entries. There are practically no limitations to the amount of data that can be added, including multimedial data, which results in better quality data. Various search options and different possibilities of database organization contribute to the efficiency of access. Dictionaries can be easily customized for specific needs of users' groups. Electronic lexicography also enables hybridization, by breaking limits between different types of language resources—for example, dictionaries, encyclopedias, term banks, lexical databases, translation tools and the like. Finally, active user involvement is possible, by enabling collaborative or community-based input to dictionaries [8].

This paper presents a data driven approach aimed at using opportunities offered by electronic lexicography, as well as various available techniques of Natural Language Processing (NLP), to develop a semi-automatic pipeline for dictionary production. The approach is focused on raw material terminology, with an emphasis on terminology related to the mining industry, as a case study, the main goal being to cover Serbian and bilingual English-Serbian terminology in the raw material domain, within a system that can be used for developing web and mobile dictionary applications. In developing this system, a data driven approach is adopted, relying on available textual, lexical and terminological resources, both in printed and electronic form. Within the development of this system, printed resources, the paper dictionaries covering raw material terminology, were subjected to systematic extensive digitisation.

In this approach, besides compiling a comprehensive multilingual lexical database of raw material terminology, lexicographic methods for automatic knowledge extraction are used, including corpus data analysis, automatic data extraction, editing and publishing extracted data in (online) dictionaries. Using extracted lexicographically relevant data (lemma lists, example sentences, collocations) as complementary resources in electronic dictionaries is known as the one-click dictionary or push-pull dictionary model, which is used, for example, in the Sketch-engine [9] for several languages, but has not yet been used for Serbian.

A similar approach to the one outlined in this paper was applied in development of the Sõnaveeb language portal of the Institute of the Estonian Language, which contains data from a number of dictionaries and termbases, with a total of 200,000 Estonian headwords with collocations, etymology, multi-word expressions, etc. The main issues to be resolved in their approach were the consistency of information, deduplication, parsing data fields containing more than one data element, moving from annotating form (e.g., italics) to annotating content (e.g., a citation) [10].

López-Úbeda et al. [11] present another interesting approach, which also combines different NLP techniques to develop a system for identification of biomedical terms in textual documents written in Spanish. The approach was applied for recognizing biomedical entities in various types of texts, including different knowledge resources (MedLine Encyclopedia, International Classification of Diseases, Unified Medical Language System, etc.). Although the tool developed within their approach has been developed for Spanish, the authors plan to expand its usability by incorporating multilingual support in the future, thus enabling it to be extrapolated to other languages.

The web and mobile applications for raw material terminology developed as a result of our approach are primarily intended for students and engineers involved in the raw material industry, as an aid in mastering terminology. They offer both English-Serbian and Serbian-English terminology, developed, inter alia, on using a comprising a variety of literature from the field of raw materials. Existing terminological dictionaries and general language dictionaries served as control dictionaries (listed in the bibliography and described in Sections 2.1 and 3.1). The developed dictionaries are not comprehensive, but rather contain basic terminology from various raw material subdomains (areas), needed to make reading professional literature easier, academic writing purposes and to improve communication among professionals in the raw material industry. In addition to core raw material terms, some technical and academic vocabulary is also introduced, that is, words that often appear in professional literature.

The developed dictionaries are not prescriptive, as they do not prescribe how the terminology "should" be systematized, but rather record the terms in use. Therefore, they feature synonyms and also record technical jargon and localisms next to standard terminology. For example, *'rotorni bager'*, namely, *'bucket wheel excavator'*, is recorded on the Serbian side together with *'glodar'*, a jargon term, literally translated as *'gnawer'*. The publication of the dictionaries as a mobile app is especially important in view of the fact that the job of an engineer dealing with raw materials usually involves frequent field work and staying in the field for prolonged periods.

Section 2 gives an overview of available resources: paper and electronic dictionaries, as well as corpora used. Section 3 outlines preparation of resources, which includes digitization of paper dictionaries, enlargement of corpora, adding domain terms to general purpose morphological e-dictionaries and extraction of bilingual lists. The process of terminology compilation, from the perspective of monolingual and bilingual extraction, a well as the web and mobile form of the dictionary are given in Section 4. The last section offers a discussion, concluding remarks and outline of future plans for improvements and application in other areas.

2. Available (Terminological) Resources

Our approach relies heavily on available resources, both in paper and electronic form, such as traditional, paper dictionaries used in raw material industry, termbases covering raw material terminology, corpora of texts from the raw material domain as well as general-purpose electronic dictionaries of Serbian. This section offers an overview of these resources.

2.1. Paper Dictionaries for Raw Material Domain

The Bureau of Mines (U.S. Department of the Interior) had pioneered efforts in mining terminology, beginning in 1918 with Fay's "Glossary of the Mining and Minerals Industry", and continuing by the 1968 publication of "A Dictionary of Mining, Minerals, and Related Terms" (DMMRT). In this 5-year project, more than 100 bureau personnel (engineers, scientists, and editors) were involved in the technical review and publication production process of the dictionary, with 28,750 terms explained by 37,180 sense definitions [12]. This dictionary has been used for several decades at the University of Belgrade Faculty of Mining and Geology (UBFMG), and it is the main dictionary covering mining terminology in English in our approach. Online version of dictionary is published on The Edumine platform that provides professional development training for people in the mining industry [13].

A multilingual "Mining dictionary: Serbo-Croatian: English: French: German: Russian" (MD), containing 16,500 terms related to underground and surface excavation, preparation of mineral raw materials, as well as rock and soil mechanics in five languages was published in 1970 [14]. This dictionary also contains terms from the fields of geology, metallurgy, electrical engineering, mathematics with computational methods, and civil engineering, to the extent they are related to mining. Each term entry has a Serbian headword, sometimes followed by synonyms, which is aligned with translations in four languages—English, French, German, and Russian. The interconnection of all five languages is given by additional indexes. Term entries do not have definitions nor usage examples. the dictionary being almost 50 years old, many terms are outdated, while some new terms are missing. This dictionary was our main source for extracting terminological equivalents in Serbian and English.

The first terminological "English-Croatian-Serbian Petroleum Dictionary" for the field of petroleum engineering [15] was followed, after 30 years, by the "English-Croatian encyclopedic dictionary of oil and gas exploration and production" [16], which is used both in Croatia and Serbia. With 12,200 definitions and 7100 terms, it contains a comprehensive vocabulary of both scientific and professional terms used by scientists, experts and students in the area of exploration and production of oil and gas, but also petroleum geology, geophysics, development deposits, drilling and equipping wells, ecology and other disciplines.

There is also a small bilingual dictionary of mineral processing [17] with 2415 translation pairs, in both directions, English to Serbian and Serbian to English, but also without definitions. Finally, a glossary of mineral processing terms with 1400 definitions in Serbian is used at the UBFMG, although it was not officially published [18].

All these dictionaries, and a number of other dictionaries, a total of 22, have been digitized for the purpose of our approach.

2.2. Digital Resources in Raw Material Domain

The development of digital resources for raw material terminology has been an ongoing activity at the UBFMG for several years now. It started with research related to the development of an ontology of mining equipment [19], in line with other research aimed at development of bilingual lexical resources [20]. The focus was then turned to development of termbases for the general field of mining engineering, and their transformation from their initial custom in-house scheme into the TermBase eXchange (TBX) Standard [21]. Another terminological resource, mostly handcrafted, was also developed to support knowledge management in specific subfields of mining engineering, such as mining equipment, mine safety and geostatistics [22]. A thesaurus of mining terminology is available online, but it is not systematically updated. Moreover the application has no new features, and it is not responsive. A modest experiment was made with developing students' vocabulary related to raw materials through flashcards and L1 in the CLIL Classroom [23], but it was not finalized with publicly available online resources.

Three digital resources already developed at UBFMG were included in our approach, two termbases, Termi [24], and GeoliSSTerm [25], and one ontology, Rudonto [26]. Termi supports development of terminological dictionaries in various fields (mathematics, computer science, raw material, library science, computational linguistics, power engineering, etc.) [27,28], and it has been selected as the most suitable resource to be used for the comprehensive multilingual lexical database of raw material terminology, while the remaining two resources have been incorporated in the dictionary production pipeline.

For systematic development of raw material terminology, textual resources, namely, bilingual libraries and corpora are also needed. Thus, articles from the scientific journal Underground Mining, published both in Serbian and English, stored in the bilingual digital library Bibliša, as one of the collections of aligned English-Serbian bi-texts [29,30], were also used in our approach.

A monolingual corpus from the mining domain was developed as part of a project related to managing mining project documentation using human language technology [31] and used within this research in the web and mobile applications.

2.3. General Purpose Morphological Dictionaries

Serbian has an extensive system of inflection and a complex agreement system that makes extraction of terminology more complicated, and thus the use of general purpose morphological dictionaries is indispensable for every lexicographic task [32].

An important lexical resource used for morphological analysis and extraction are the comprehensive electronic morphological dictionaries for Serbian (SrpMD) of simple- and multi-word units, covering general lexica, proper names, encyclopedic knowledge and terminology from a number of domains [33], with nearly 200.000 lexical entries. SrpMD entries include both a lemma and inflected forms supplied by grammatical information, semantic markers, domain information and relations of several types: derivational, lexical variation, component relations (between single words and terminological phrases).

For example, lexical entry *'rudar'* (miner, person engaged in mining, a worker in a mine) contains information related to part of speech: *'N'* (noun), morphological class *'N2'*, semantic tag *'+Hum'* (human), domain *'DOM = mining'*. Its inflected forms are: *'rudar'* (ms1v), *'rudara'* (mp2v:ms2v:ms4v:mw2v:mw4v), *'rudare'* (mp4v:ms5v), *'rudari'* (mp1v:mp5v), *'rudarima'* (mp3v:mp6v:mp7v), *'rudarom'* (ms6v), *'rudaru'* (ms3v:ms7v) where brackets show grammatical information: *'m'*—masculin, *'s'*—singular, *'p'*—plural, *'1–7'*—cases, *'v'*—animate.

The entry *'rudar'* is also related to the relational adjective *'rudarski'*, and appears as a component of several terminological phrases, for example, rudar na okresivanju (ripper), rudar na uglju (collier), rudar-podgrađivač (timberman), and so forth.

Over the past years, more entries related to raw material were added to SrpMD, which initially contained more than 3000 simple-word entries and 2000 multi-word entries from the raw material domain. The number of their morphological forms recorded in this resource is significantly larger. The simple-word forms pertaining to raw material terminology that have been processed and included in SrpMD [34] enabled further extraction of related terminological phrases according to the methodology described in [19]. Namely, for extraction to be effective, it is very important that the domain is relatively well covered with simple domain-specific words.

3. Resource Preparation

Preparation of resources is aimed at expanding and enriching available digital resources. These activities are not to be understood as one-time only activities, as each of them can be repeated periodically, when new opportunities for resource enrichment appear.

3.1. Dictionary (Retro)Digitisation

In order to expand and enrich the available digital resources, a number of paper dictionaries were digitised in the preparatory phase. After scanning, OCR and transformation to MS Word, with preservation of formats (bold, italic), manual correction was performed. The Word documents were then parsed, by a parsing procedure that was fine-tuned for each dictionary, according to its structure. Parsed data were finally transformed to structured formats: excel and xml, before being imported to the internal relational database. The procedure will be illustrated on one multilingual dictionary (MD) and one monolingual dictionary (DMMRT).

The digitisation and parsing of MD produced 16,491 term entries (examples of term entries are given in (Figure 1), where Serbian terms were aligned with one or more English term equivalents (the remaining 3 languages were also stored in the database, but they were not used in this approach).

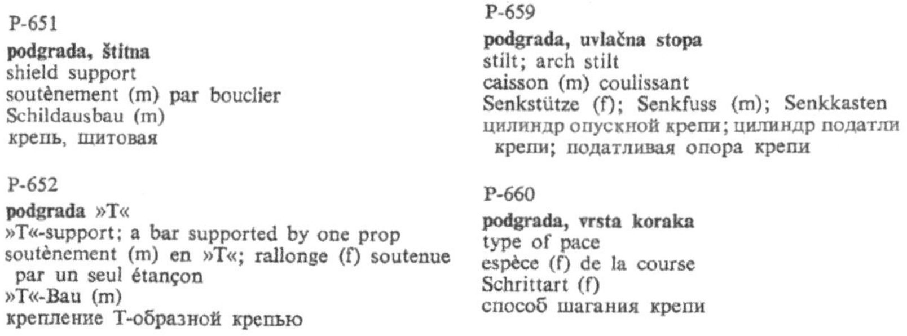

P-651
podgrada, štitna
shield support
soutènement (m) par bouclier
Schildausbau (m)
крепь, шитовая

P-652
podgrada »T«
»T«-support; a bar supported by one prop
soutènement (m) en »T«; rallonge (f) soutenue par un seul étançon
»T«-Bau (m)
крепление Т-образной крепью

P-659
podgrada, uvlačna stopa
stilt; arch stilt
caisson (m) coulissant
Senkstütze (f); Senkfuss (m); Senkkasten
цилиндр опускной крепи; цилиндр податли крепи; податливая опора крепи

P-660
podgrada, vrsta koraka
type of pace
espèce (f) de la course
Schrittart (f)
способ шагания крепи

Figure 1. Examples of scanned Mining dictionary entries.

The majority of dictionary entries (15,016) contained only one Serbian term, but there were 1355 entries with two terms, and 120 with 3–5 terms, resulting in a total of 18,092 Serbian terms, of which 16,916 distinct. As to the English part of the dictionary, there were 13,163 entries with one term, 2553 with two terms and 775 with 3–8 terms, resulting in a total of 20,878 English terms, of which 17,774 distinct.

Raw material terminology, akin to general technical terminology, contains a large number of multi-component terms. In the dataset obtained from the dictionary 23% of English entries are single word terms, 50% are two-component terms, 18% have three components and the remaining 9% have four or more. As for Serbian entries, 22% are one-component terms, 47% have two components, 17% have three, and the remaining 14% have four or more. The majority of English multi-compound terms are noun compounds. These linguistic constructions are most often composed of two or more nouns. for example, *'coal waste'—'jalovina'*, *'waste dump'—'odlagalište jalovine'*, *'gas pressure'—'pritisak gasa'*. However, they can also contain three, four or more nouns, for example, *'gas protection apparatus'—'lična zaštitna sredstva od gasova'*, *'mud circulation pressure hose'—'isplačno crevo'*.

Given the frequency of multi-component terms, an analysis of translational equivalents in English and Serbian was performed in terms of the number of their components. It was found that in 20% of cases both translational equivalents have one component, in 31% of cases both have two components, in 15% of cases the Serbian term has one component more than the English term, while in 13% of cases the English term has one component more, in 5% of cases the Serbian term has two components more, and in 3% of cases English has two components more. All other cases cover the remaining 13% of cases.

Entries in DMMRT have one or more senses per each term, described by a definition, and labeled by small letters *a*, *b*, *c*,. . . , *u*. Each individual sense can be related to one or more other terms in the dictionary, and it can be followed by its bibliographic source. Digitization of DMMRT yielded 28,757 terms with a total of 37188 sense definitions, where 24,115 terms have only one sense, 2942 have 2, 890 have 3, 641 have 4–6, 139 have 7–10, and 34 have 11–21. The most polysemous word is *'head'* with 21 senses, followed by *'drift'* and *'bottom'* with 20 senses. Types of relations between entries can be: See (4090), See also (3983), CF (compare, 1824), Ant (antonym, 20), Etymol. (etymology, 130), Syn: or syn.(synonym, 2532), Abbrev. (abbreviation, 77), etc. Figure 2) presents the entry *'accessory plate'* with five senses, marked by letters a-e. Two senses (a and e) are related to other dictionary terms (a to *'quartz wedge'* by CF, and e to three synonyms and two other terms by CF), and two senses (b and c) are followed by their source (Pryor).

accessory plate

a. The quartz wedge inserted in the microscope substage above the polarizer in order to estimate birefringence and to determine optical sign of uniaxial minerals. CF: quartz wedge b. The selenite plate that gives the sensitive tint of a specimen between crossed nicois. *Pryor, 3* c. The mica plate that retards yellow light. *Pryor, 3* d. In polarized-light microscopy, an optical device that may be inserted into the light train to alter light interference alter passage through, or reflection by, a crystalline material; e.g., quartz wedge, mica plate, gypsum plate, or Bertrand lens, e. In polarized-light microscopy, an optical compensator that may be inserted into the light train to alter birefringence after light passage through or reflection by an anisotropic material; e.g., quartz wedge, mica plate, gypsum plate, or Berek compensator. Syn: gips plate; glimmer plate; compensator. CF: Berek compensator; gypsum plate.

Figure 2. An example of scanned entry from DMMRT.

As to the components of the terms in DMMRT, 37% of the total terms are single word terms, 50% are two-component terms, 10% have three components and the remaining 3% have 4–7 components. Comparison with the English part of MD shows a similar pattern, as the percentage of two-component words is equal, while MD has 14% less one-component terms.

Additional 19 dictionaries from the raw material and related domains were digitized, parsed and stored in the database, adding 63,571 new entries. Five monolingual English dictionaries from the mining domain produced 5933 entries, three bilingual mining English-Serbian dictionaries produced 24,049 entries, three monolingual English dictionaries covering terminology from the mine safety domain contributed with 655 entries, and an English-Serbian dictionary of terminology in the field of waste management yielded 1968 entries. Dictionaries from related domains were also included, namely four English dictionaries producing 21,448 entries and three bilingual dictionaries producing 9518 entries.

One of the observations, even before this research started, was that several terms in paper dictionaries are not in use anymore. That observation initiated frequency calculation of Serbian terms in the mining corpus. Frequency in the corpus and the number of dictionaries that attest a term were the main criteria for post editing priority of the term.

Entries from all digitized dictionaries were stored in the same database, but in different structures, which correspond to their original data schema, and with reference to the original source. All of the structures can, in general, be mapped to the union of the structures of the two dictionaries presented in more detail, MD and DMMRT. Thus, a terminological entry in the common database can consist of a headword (list), rarely part-of-speech, equivalent(s) in other language(s), usually one, but sometimes more, labeled senses that include definitions, occasionally synonyms and abbreviations, links to other entries, bibliography, rarely specific domain.

3.2. Corpora Enlargement

The monolingual corpus of texts from the mining domain and related research work, which comprised 172 documents (in Serbian) with 2.7 million words in first release [31], was subsequently enlarged with 63 documents. The current version has 4.1 million words, covering project documentation (26%), legislation (11%), doctoral dissertations (31%), textbooks and other mining literature (32%).

The bilingual corpus of texts aligned on the sentence level was produced from the bilingual digital library Bibliša. The initial set of 55 documents containing 4831 aligned Serbian-English sentences [29] was enlarged with 44 new documents containing 12,657 aligned sentences from the raw material and energy domains.

The crucial linguistic preprocessing steps within corpora enlargement are part-of-speech tagging and lemmatization. Part-of-speech tagging represents an automatic text annotation process in which words or tokens are marked by part of speech tags, which typically correspond to the main syntactic categories in a language (e.g., noun, verb). Lemmatization is the process by which inflected forms of a lexeme are grouped together under a base dictionary form. The Serbian corpus and the Serbian part of the bilingual corpus are tagged and lemmatized using a customised tagger [35], while the English part of the bilingual corpus is tagged by Treetagger [36,37].

Texts included in corpora are also processed using electronic dictionaries and local grammars. It is important to note that text processing and related mining vocabulary expansion is an iterative process. Namely, among other tasks, corpora are used for extraction of mining terminology, definitions and usage examples by applying different methods and tools.

3.3. Adding New Serbian Terms to General Lexica Dictionaries

Terminology from digitized dictionaries of raw material terminology in Serbian was checked by SrpMD and the corpus from the mining domain, for possible adding to SrpMD. We will illustrate this procedure by the results obtained from MD. The Serbian part of MD that contains headwords was transformed into a text, which was then analysed by SrpMD. Out of 12,655 different single words found in the text produced from the dictionary, 9758 were recognized by SrpMD. Among the 2897 (23%) that were not recognised, there were some acronyms (e.g., *'pH'*, *'RR'*, *'LD'*, *'TV'*), names (e.g., *'Western'*, *'Bets'*, *'Reni'*), archaisms (e.g., *'abanje'* instead of *'habanje'* (wear and tear), *'bolcn'* instead of *'zavrtanj'* (screw), etc.), as well as some OCR errors (despite manual check-up). Based on this analysis, a set of candidates for new entries into SrpMD were prepared (e.g., *'degazacija'* (degassing), *'eksploatabilan'* (exploitable), *'sabirnik'* (busbar), etc.). Each candidate was further checked against the mining corpus, and if the result (basically, its frequency) was satisfactory, it was added to the SrpMD.

The same procedure was applied to other dictionaries with Serbian entries. While the comprehensive terminological dictionaries (such as MD) contained a lot of simple words that were missing in SrpMD, smaller dictionaries, as expected, included frequently used terms that were mostly already in SrpMD. Thus, for example, in Electropedia 13% of words were not recognized by SrpMD, while in the Serbian part of the English-Serbian dictionary of terminology in the field of waste management 6% of words were not recognized. In all other dictionaries the percentage of unrecognized words was between 3%–5%, but whether they would be included into SrpMD depended on their frequency in the mining corpus.

Besides the digitized dictionaries, the Serbian corpus and the Serbian part of the bilingual corpus from the mining domain were yet another source of new raw material domain terms that did not exist in SrpMD. Extraction of simple words was relatively simple, namely, words that were not recognized by SrpMD were scrutinized, and if frequent enough, they became candidates for being added to SrpMD. Besides, less than 4% of words in the monolingual mining corpus were unrecognised by SrpMD, where approximately 1.3% out of these 4% were proper candidates to be added to SrpMD, the remaining unrecognized words being variables from equations (0.7%), acronyms (1%), low frequency (hapax and typos—0.5%), foreign names and words (0.5%).

However, when it comes to terms in the form of terminological phrases, their extraction from corpora becomes much more complicated. Automatic extraction of term candidates for Serbian relies on a procedure presented in [30,34]. Essentially, it is based on detecting words in corpora that follow one of the 23 specific syntactic patterns, most frequent for noun terms (AN adjective-noun, NNg noun-noun in genitive case, AAN, ...). The first step in this task is to recognise and extract Serbian terminological phrases from the corpus using syntactic patterns, and calculate their frequency. Frequency was the main parameter for determining the rank of a terminological phrase as a candidate for processing for SrpMD. However, other measures of association, such as T-Score, Keyness, Log-likelihood, were also used, as described in detail in [30]. The task then proceeds by lemmatization of candidate terminological phrases, disambiguation for terminological phrases where more lemmas can be produced, and ends by production of the final lemma, which enables production of all inflected forms for each terminological phrase.

As in the case of single terms, frequency for terminological phrases was also calculated for each single-word component of the phrase, but for its lemma, not for the exact inflected form. Having in mind free word order in terminological phrases we were looking for a measure more loose than exact match. For each terminological phrase the following information is stored: minimum, average and maximum frequency of its components, number of "known" components-words recognized by SrpMD. Frequency in the corpus and the number of dictionaries that attest a term are the main criteria for post editing priority of the term.

For this paper, extraction of Serbian terminological phrases was performed with a frequency threshold of 10, and 12,632 candidate phrases were produced in lemmatized form. Frequency of each terminological phrase was calculated as the sum of frequencies of all its inflected forms. For example, *'kvalitet uglja'* (coal quality) has a frequency of 1110 as a sum of frequencies of its forms: *'kvalitet uglja'* (172), *'kvaliteta uglja'* (587), *'kvalitetom uglja'* (284), *'kvalitetu uglja'* (53), *'kvalitete uglja'* (8), *'kvaliteti uglja'* (2), *'kvalitetima uglja'* (4). Six most productive patterns, which produced 92% of candidates, are listed with examples and their frequencies:

- NNgi (32%), N2X—a noun followed by a word that does not inflect in the terminological phrase. Usually this word is a noun in the genitive or in the instrumental case; examples are *'kvalitet uglja'* (coal quality—1110), *'sistem upravljanja'* (management system—902), *'procena rizika'* (risk assessment—514).
- AN (29%), AXN—an adjective followed by a noun; the adjective and the noun have to agree in all four grammatical categories; examples are *'površinski kop'* (open pit—5738), *'ugljeni sloj'* (coal seam—1686), *'rudarski projekt'* (mining project—1412).
- NprepNp (11%), N4X—a noun followed by two words that do not inflect in the terminological phrase where these word form a prepositional phrase; examples are *'zdravlje na radu'* (occupational health—1323), *'čvrstoća na smicanje'* (shear strength—270), *'transporter sa trakom'* (belt transporter—240).
- N-N (10%), NXN—a noun followed by a noun that agrees with it in number and case, where the separator can be a hyphen; examples are *'gas-lift'* (197), *'blok dijagram'* (block diagram—192), *'bager vedričar'* (bucket excavator—174). This class had the largest number of recognized phrases for rejection, that is, those whose slightly different lemmas were already captured by another pattern, and this pattern should thus be placed with some lower priority in disambiguation.
- X-N (6%), 2XN—a noun preceded by a word that does not inflect in the terminological phrase. Usually it is a word that is used only in one or few terminological phrases, a prefix or an adverb derived from an adjective, while the separator can be a hyphen; examples are *'bto sistem'* (bto system—1728), *'pm preduzeće'* (pm company—373), *'y-osa'* (y-axis—19).
- NNgiNgi (4%), N4X—a noun followed by two words that do not inflect in the terminological phrase where these two words are adjectives/nouns in the genitive or instrumental case; examples are *'zaštita životne sredine'* (environment protection—668), *'eksploatacija mineralnih sirovina'* (mineral resource exploitation—228), *'efekat staklene bašte'* (greenhouse effect—109).

Evaluation follows, where the following is checked: is the extracted candidate a terminological phrase, which domain (mining, technical, etc.) and possibly subdomain it belongs to. If the domain or subdomain are identified, the appropriate semantic markers are assigned to the terminological phrase. After the evaluation process, all correctly evaluated terminological phrases were prepared for insertion into the terminological database Termi.

3.4. Adding Bilingual Terms

Bilingual lists of terms were considered a valuable resource in our approach, and they were generated from two sources, namely, by retrieval from the bilingual MD and by extraction from the aligned bilingual corpus.

Term entries from MD were parsed and only those that were confirmed by the mining corpus (monolingual or bilingual) were selected. As mentioned before, one term entry can comprise more terms (single or multi word) and confirmation for each term was looked for.

A total of 10,059 term entries from MD were retrieved, with sets of English terms aligned with sets of Serbian terms. The majority of them were subsequently marked by domain (24 different), subdomain (15) and semantic markers (35) as mentioned in Section 3.1. All markers used are subsets of markers—data category values in srpMD.

Bilingual terminology was extracted from the aligned bilingual domain corpus described in Section 3.2 using terminology extractors for Serbian and English, and Bilte [38]), a tool for chunk alignment [39,40]. The method combines the approach with existing domain terminology lexicons with term extraction tools. For English, FlexiTerm [41] was used with threshold 3 and TermSuite [42] with threshold 4, based on the experience from other domains and the fact that they use different linguistic filtering. A total of 8456 term candidates for English were selected. For Serbian, the same shallow parser was used as in the case of monolingual extraction (Section 3.3), as well as the same calculation of termhood, a frequency-based measure, which qualified 7825 candidates as terms.

Monolingual lists of extracted terms were further expanded by terms retrieved from digitized dictionaries yielding 94,539 English terms and 48,096 Serbian terms. Some terms were found in both datasets: extracted from text and retrieved from dictionaries, namely, a total of 2285 English and 308 Serbian terms.

The GIZA++ [43] and Moses toolkit [44] for statistical machine translation (SMT) were used for word alignment. Aligned chunks, presented in the so-called phrase table, are obtained as output from Moses, together with their phrase translation scores. After pruning the phrase table with the threshold probability of 0.85, the remaining chunks were lemmatized and further filtered to select those in which both parts of the pair contain a candidate term from the raw material domain. More details about options and the procedure are available in [40]. The output of this phase contained 8202 Serbian-English pairs as term candidates whose English part was confirmed and 3605 where both language parts were confirmed. In the first step, candidates that were found in digitized dictionaries, or were already assessed as terms, were automatically confirmed, but candidate pairs had to be inspected manually, which yielded a list of 2737 term pairs. General terms, such as, *'red' (row)*, *'kompozicija' (composition)*, *'din' (dinar)*, *'minimalan' (minimum)*, *'izvor informacija' (source of the information)*, ... were excluded, as well as those wrongly aligned, such as: *'naftovod' (pipeline oil)*, *'mreža' (telephone network)*, *'deponija' (deposit)*, *'oblik poklopca' (shape of the cover)*, ... A wider set of terms will be evaluated in the near future.

For evaluation of bilingual candidates, besides frequencies for single terms, we have also used a heuristic for evaluating terminological phrases based on the following observations. The last noun in English noun compounds, which represent the majority of English terminological phrases, as a rule, is the head word carrying the basic meaning, while the preceding nouns are narrowing this meaning, that is, behaving like adjectives. The meaning of a noun compound in English thus flows from right to left, but the Serbian translational equivalent cannot be formed analogously, namely, by a sequence of corresponding Serbian nouns. Thus, within the analysis, the most frequent constructions used as Serbian translational equivalents for English noun+noun compound were determined:

- noun + noun in the genitive (e.g., *'coal mining'-'eksploatacija uglja'*)
- adjective + noun (*'waste water'-'otpadna voda'*)
- noun + prepositional phrase (*'belt conveyor'-'transporter sa trakom'*)
- paraphrase (*'crusher stower'-'mašina za drobljenje i pneumatsko zasipanje'*)
- one-word name (*'crushing machine'-'drobilica'*).

This heuristic was used to select the most promising candidates among the extracted bilingual terminological phrases.

As in the case of multilingual terms and terminological phrases, after the evaluation process, all correctly evaluated bilingual terms were prepared for insertion into the terminological database Termi. So far, more than 3000 term-to-term pairs were inserted. In this process they were merged to form synonymous sets (synsets) by using information from existing dictionaries and simple rules, such as: if two English terms are translated by the same Serbian term they are candidates for synonyms.

4. Terminology Aggregation and Presentation

4.1. Data Integration Procedure—The Pipeline

The main goal of our approach is to merge and link all available terms in the raw material domain into one lexicon structure, within the terminological database Termi and as linguistic linked data available via SPARQL endpoint, in the first place by aligning as much as possible term entries from dictionaries and other resources covering raw material domain terminology. Besides the aim of aggregating terms from different resources, one of the reasons for alignment of terms from multiple dictionaries (paper and electronic) was to assess term usage, which determines its importance for raw material terminology. On the other hand, alignment of terms with SrpMD was necessary, since these dictionaries are a base resource for lemmatization and multiword term extraction. Since SrpMD are already in the lexical database Leximirka [32], developed and managed by the same research team, this type of alignment was possible.

Figure 3 presents an outline of the pipeline for termbase population, which starts with collecting and preparing research papers, project documentation, and textbooks in Serbian for the monolingual corpus and aligning English-Serbian texts for the bilingual parallel corpus. Also, paper dictionaries, both monolingual and bilingual are digitized, parsed and stored in an auxiliary database as structured data in XML format.

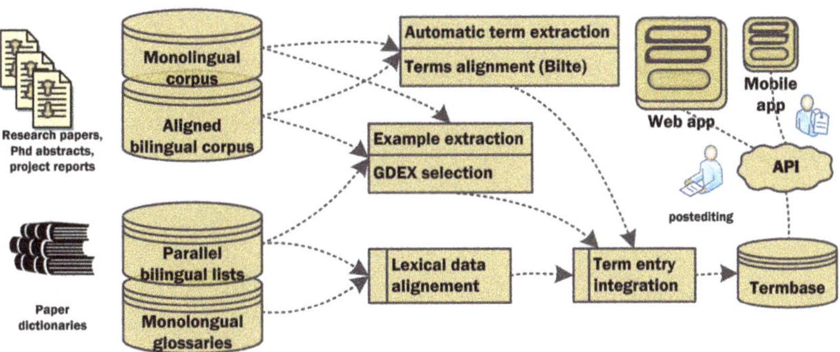

Figure 3. The pipeline for terminology compilation (termbase population).

Compiled resources also comprise monolingual lists derived from all available resources, interlinked with their source entries, for example Serbian list from Serbian monolingual dictionaries and Serbian part of bilingual dictionaries. Translation equivalents are retrieved from bilingual dictionaries and within the word alignment phase (more in Section 4.2), keeping again information about the original dictionary source.

Extracted terms were also subject to a labeling procedure, which we will illustrate here on the example of MD. Out of 16,491 entries obtained from MD, 12,018 (73%) were manually classified and markers for domain and subdomain, as well as semantic labels, were assigned to them. The remaining 4473 (27%) unclassified entries included words from general lexica and some rarely used terms. The classified entries are mostly from the mining domain, more precisely, there are 4793 (40%) entries common for different areas of mining. The basic vocabulary from related domains is also included, for example, 2398 (20%) entries related to geology, hydrogeology and geography, 860 (7%) entries related to transport, rock mechanics, surveying, environment protection, safety, construction, transport and electrical engineering, while 3082 (26%) entries belong to the general technical terminology. There are also entries from basic science, for example, 885 (7%) terms related to biology, chemistry, mathematics, informatics and physics.

Among entries from the mining domain, those related to a specific subdiscipline of mining were identified by mining experts, and marked by a subdomain marker, as for example, entries related to mineral processing (251), transport (243), or underground mining (469). Additional semantic labels were also assigned, for example, material (699), device (536), machine (384), mineral (313), facility (288), instrument (279), etc.

The part-of-speech was semi-automatically assigned, where only 40 entries were marked as adjectives, 250 as verbs, and all other as nouns.

Lexical entry alignment with DMMRT is performed using terms on the English side of the MD. Since one English term can have several senses, such alignments are marked for manual filtering. An indicator is used for status: automatic relation or manually evaluated.

A terminological dictionary must accompany each entry with a scientifically and lexicographically correct definition [45]. There are very few such dictionaries in the Serbian language, as most of the published Serbian terminological dictionaries are only translational (bilingual or multilingual). An ongoing activity is the adaptation of English definitions, which are the most comprehensive in DMMRT, to Serbian, in the post-editing phase, where priority is given to the most frequent terms, both in the corpora and in the dictionaries.

Finally, candidates are harmonised and assembled to the microstructure of the lexical database Termi, which consists of a headword, synonyms, abbreviations, definition, for each language, bibliographic source and possibility to include illustration and other external content. Term entries in Termi are organised into a hierarchical structure, and additional relations between entries are envisaged, but still not implemented. Automatic hierarchical positioning was based on subdomain and semantic markers, but it is subject to repositioning in the post-editing phase.

Information integration beyond the level of individual dictionaries and across the language resource community has become an important concern, and the most promising technology to achieve this goal is to adopt the Linked (Open) Data (LOD) paradigm for publishing lexical resources, that is, to use URIs for unambiguously identifying lexical entries, their components and their relations in the web of data—to make lexical datasets accessible via http(s), to publish them in accordance with W3C-standards such as RDF and SPARQL, and to provide links between lexical data sets and with other LOD resources [46].

In our research we were also aiming at compatibility with the Linked Data approach, using its set of design principles for sharing machine-readable interlinked data on the Web. This vision of globally accessible and linked data on the internet is based on RDF standards of the semantic web, using RDF serialisation for data representation. To that end, our approach envisages export of lexical database data in RDF that is compliant with the *The OntoLex Lemon Lexicography Module* [47], lexicog [48], as an extension of Lexicon Model for Ontologies (lemon) [49,50]. This is also in line with activities within NexusLinguarum COST action [51], which promotes synergies across Europe between linguists, computer scientists, terminologists, language professionals, and other stakeholders in industry and society, in order to investigate and extend the area of linguistic data science. An example of RDF export is presented in Figure 4 followed by the Turtle RDF Syntax [52] to illustrate the use of the model.

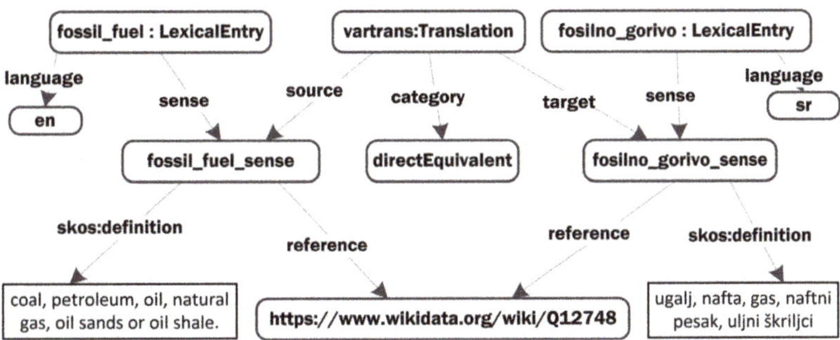

Figure 4. The graph for the translation of lexical entries: *'fossil fuel'-'fosilno gorivo'*).

```
:fossil_fuel a ontolex:LexicalEntry;
   dct:language <http://lexvo.org/id/iso639-1/en> ;
   lexinfo:partOfSpeech lexinfo:noun;
   ontolex:lexicalForm :fossil_fuel-form;
   ontolex:sense :fossil_fuel_sense.
:fossil_fuel-form a ontolex:Form;
   ontolex:writtenRep "fossil fuel"@en.
:fossil_fuel_sense  skos:definition "coal, oil, gas, oil sands or oil shale"@en;
   ontolex:reference <https://dbpedia.org/page/Fossil_fuel>;
   ontolex:reference <https://www.wikidata.org/wiki/Q12748>;
   ontolex:reference <http://eurovoc.europa.eu/6045>.

:fosilno_gorivo a ontolex:LexicalEntry;
   dct:language <http://id.loc.gov/vocabulary/iso639-1/sr> ;
   lexinfo:partOfSpeech lexinfo:noun;
   ontolex:lexicalForm :fosilno_gorivo-form;
   ontolex:sense :fosilno_gorivo_sense.
:fosilno_gorivo-form a ontolex:Form;
   ontolex:writtenRep "fosilno gorivo"@sr.
:fosilno_gorivo_sense skos:definition "ugalj, nafta, gas, naftni pesak ili
   uljni škriljci"@sr;
   ontolex:reference <https://www.wikidata.org/wiki/Q12748>.

:trans_fossil_fuel_sense-fosilno_gorivo_sense a vartrans:Translation;
       vartrans:source :fossil_fuel_sense;
       vartrans:target :fosilno_gorivo_sense;
       vartrans:category
           <http://purl.org/net/translation-categories#directEquivalent>.
```

Further details related to the above example, namely, the novel module for frequency, attestation and corpus information (FrAC) [53] is described in the next section.

4.2. Dictionary Examples and Frequencies

None of the dictionaries we have used contain examples of term usage. Our intention was to select actual terms that can be found in domain texts and to link usage samples to both monolingual and bilingual terms entries. Previous (and actual) practice in Serbian lexicography has relied on retrieving example candidates and definitions manually from different online sources and printed material (over a number of years), but it is evident that a more systematic and corpus-evidence-based approach was needed.

A method for the selection of good examples for Serbian terms was developed based on a feature extraction web services and knowledge retrieved from SASA Dictionary as the Gold Standard for Good Dictionary Examples (GDEX) for Serbian [54]. The method is based on a detailed analysis of various lexical and syntactic characteristics of examples in published dictionaries. The initial set of functions was inspired by a similar approach for other languages. The distribution of the characteristics of examples from this corpus is compared with the characteristics of the distribution of the sample sentences extracted from the corpus that contains different texts. The approach was adapted to work also for English and to be applied for bilingual aligned sentences. For ranking, we have used a weighted score derived from lexical features (e.g., sentence length, number of all no space chars, digits, weird chars, commas, full stops, punctuation, number of all tokens, average token length, max token length, sentences between 15 and 40 tokens, . . .), word-based features (e.g., number of words, capitalised words, . . .) and other features (e.g., average frequency in corpus, number of stop words, proper names, pronouns). New features were introduced for bilingual examples, for example, difference in sentence length measured in words, where examples in which a sentence in one language is short and in the other language long are avoided. An example containing terms as key words in context in English and Serbian, sentence examples and calculated features is:

```
109867|7.2011.60.8|7.2011.60.8_n44|Fossil fuel|Fosilno gorivo|Carbon emissions
from sources other than fossil fuel combustion are now incorporated in the
National Footprint Accounts.|Emisije ugljenika iz drugih izvora, ne samo iz
sagorevanja fosilnih goriva sada su ubeležene u Izveštaje o nacionalnoj stopi
emisije zagadenja.|120|104|0|37|0|1|1|True|18|5.778|12|True|True|True|True|17|
6.0588|12|2|3|0.0|7|145|124|0|52|1|1|3|False|23|5.392|12|True|True|True|True|
20|5.5|11|1|1|10955.428|7
```

For entries with no examples in the bilingual corpus, monolingual examples were extracted from the Serbian mining corpus. Apart from offering preselected examples, it is important to enable the user to browse the concordances for a lemma, as well as syntactic patterns, as presented in the next section in Figure 5.

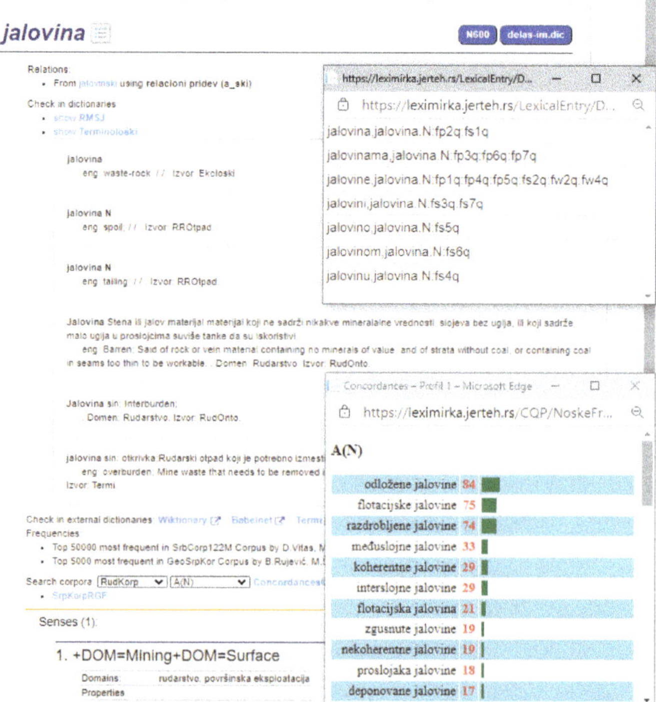

Figure 5. The Leximirka app for lexical database management.

Relative frequency (normalized per million) is assigned to terms from the mining corpus (as domain specific) and for the corpus of standard Serbian (as reference), in order to calculate the so-called keyness score, which is expected to represent the extent of the frequency difference.

Frequency information is a crucial component in human language technology, so the FrAC module includes terminology to capture such information, in order to facilitate sharing and utilising this valued information [53]. Sketch engine API [55,56] is used for calculation of frequencies, for word-sketch retrieval with collocations and for thesaurus with related words association measures (Statistics used in the Sketch Engine [57,58]). The Python script prepared in the form of a jupyter notebook was published at github [57]. Current work of the Ontolex group is focused on modeling word embeddings, collocations and similar words and we will add this feature when it becomes stable. An example of ontolex-lemon frequency and attestation snippet is:

```
# subproperty definition for frequency in mining corpus
:rudkorFrequency rdfs:subClassOf frac:CorpusFrequency .
:rudkorFrequency rdfs:subClassOf [
   a owl:Restriction ;
   owl:onProperty frac:corpus ;
   owl:hasValue <https://app.sketchengine.eu/#
      dashboard?corpname=user%2FAleksandraTomasevic%2Frudkor>] .
# frequency assessment (in mining corpus)
:fosilno_gorivo frac:frequency [
   a :rudkorFrequency;
   rdf:value "38"^^xsd:int].

# usage examples as attestations
:fosilno_gorivo frac:attestation attestation_1324567;
attestation_1324567 a frac:Attestation ;
   cito:hasCitedEntity   <https://app.sketchengine.eu/#
      dashboard?corpname=user%2FAleksandraTomasevic%2Frudkor> ;
   rdfs:comment "Dokument 31, DK_Monitoring u zivotnoj sredini" ;
   frac:locus :locus_2415677;
  frac:quotation "Koncentracija zagađujućih supstanci, posebno
   onih koje se izdvajaju sagorevanjem fosilnih goriva, varira
   u odnosu na godišnje doba (leto, zima)." .
:locus_2415677 a :Occurrence ;
   nif:beginIndex 80 ;
   nif:endIndex 96.
```

We have just started using VocBench, a web-based, multilingual, collaborative development platform for managing Ontolex-lemon lexicons among other RDF datasets [59], for publishing terminology as RDF data, in order to meet the needs of semantic web and linked data environments. VocBench is an open source web platform for collaborative development of datasets in compliance with Semantic Web standards, offering a general-purpose collaborative environment for development of any type of RDF dataset (with dedicated facilities for ontologies, thesauri and lexicons), including editing capabilities and managing SPARQL endpoint [60]. The system is able to interact with standard technologies in the RDF/Linked Data world, with the possibility to surf linked open data on the Web, access SPARQL endpoints, resolve RDF descriptions through HTTP URIs, and so forth, as well to import/export data through standard Graph Store APIs and the like.

4.3. The Web and Mobile App

The application for management of Serbian morphological dictionaries, including the evaluation of automatically extracted term candidates used in this approach is Leximirka [61]. Figure 5 presents a web page with term entry *'jalovina'*, where the user can see (1) inflected forms with grammatical categories, (2) inflectional class (*'N600'*) and dictionary (*'delas-im.dic'*); (3) dictionary entries from other dictionaries (digitized and digitally born) grouped by dictionary type (descriptive, terminological, bilingual); (4) related entries (e.g., relational adjectives *'jalovinski'*), lexical variants, derived terms; (5) corpus frequencies; (6) corpus selection with links to concordances and frequency histograms for simple lemma query or predefined syntactic patterns (in figure pattern AN where N is the headword *'jalovina'*), (7) one or more senses with semantic and domain markers.

An important feature of this system is the possibility to insert a formula in the definition, which is often necessary to precisely define a concept. The Figure 6 presents a part of the screen with a latex form of definition and its preview on the same panel. The JavaScript display engine for mathematics MathJax [62,63] that works in all browsers is used in the web application, and KaTeX [64,65] for formula rendering in the mobile application.

Parametar koji karakteriše otpornost na usitnjavanje određene čvrste materije. Određuje se pomoću laboratorijskog mlina sa šipkama ili kuglama po Bondovoj metodi. U industrijskim uslovima on se određuje po jednačini: $W_{i}=W_{t}\times \frac{\sqrt{F}}{\sqrt{F}-\sqrt{P}}\times \sqrt{\frac{P}{100}}$. Koristi se za proračun snage pogonskih motora i izbor drobilica i mlinova za usitnjavanje.
Parametar koji karakteriše otpornost na usitnjavanje određene čvrste materije. Određuje se pomoću laboratorijskog mlina sa šipkama ili kuglama po Bondovoj metodi. U industrijskim uslovima on se određuje po jednačini:
$W_i = W_t \times \frac{\sqrt{F}}{\sqrt{F}-\sqrt{P}} \times \sqrt{\frac{P}{100}}$. Koristi se za proračun snage pogonskih motora i izbor drobilica i mlinova za usitnjavanje.

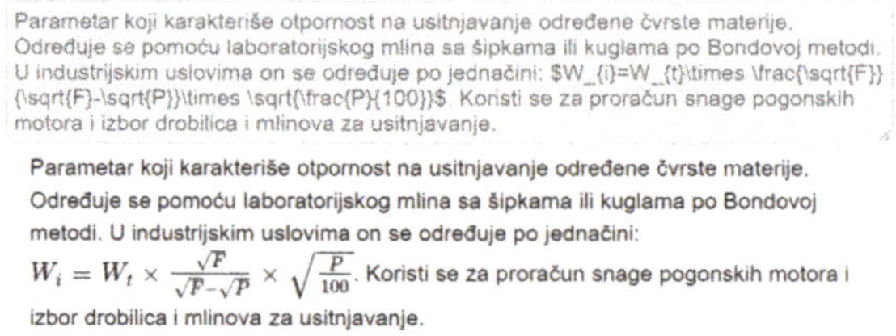

Figure 6. Formula editing and preview in term entry.

The mobile application allows the user to search for a Serbian or English term, where the query is submitted to the Termi API and a list of entries is retrieved, with a further possibility to request examples for selected entries. Figure 7 presents screenshots of mobile and web applications.

Besides for search, browse and the described export, the application can also be used for preparation of a dataset for Lexonomy [66,67]. Figure 8 presents a panel for term entry editing, which is connected with the Sketch-engine and enables retrieval of examples from a related corpus, in our case the corpus from the mining domain.

Bondov indeks
Definition:
Synonyms:
Koeficijent meljivosti
Abbreviation:
Broader terms:
priprema mineralnih sirovina; rudarsko inženjerstvo
Bibliography:
Terminološki rečnik iz pripreme mineralnih sirovina, Lazić Predrag i dr., 2020
Bond index
Definition:
A parameter that characterizes the resistance to comminution of a certain solid. It is determined using a laboratory mill with rods or balls according to the Bond method. In industrial conditions, it is determined by the equation: $W_i = W_t \times \frac{\sqrt{F}}{\sqrt{F}-\sqrt{P}} \times \sqrt{\frac{P}{100}}$. It is used to calculate the power of drive motors and to select crushers and grinding mills.
Synonyms:
Work index
Usage examples
https://termi.rgf.bg.ac.rs/D...
Prelistavanje
Pretraga
Ažuriranje
Bibliografija
Import iz excela
Profil
Pomoć
bondov indeks
Biblisha NaRA
Termin:
bondov indeks
Definicija:
Parametar koji karakteriše otpornost na usitnjavanje određene čvrste materije. Određuje se pomoću laboratorijskog mlina sa šipkama ili kuglama po Bondovoj metodi. U industrijskim uslovima on se određuje po jednačini: $W_i = W_t \times \frac{\sqrt{F}}{\sqrt{F}-\sqrt{P}} \times \sqrt{\frac{P}{100}}$. Koristi se za proračun snage pogonskih motora i izbor drobilica i mlinova za usitnjavanje.
Sinonimi:
koeficijent meljivosti
Skraćenice:
Nadređeni koncept:
priprema mineralnih sirovina rudarsko inženjerstvo
Bibliografija :
Terminološki rečnik iz pripreme mineralnih sirovina, Lazić Predrag i dr., 2020
Termin:
bond index
Definicija:
Sinonimi:
work index
Pretraži korpuse:
GeoSrpKor
Konkordance
Frekvencije oblika

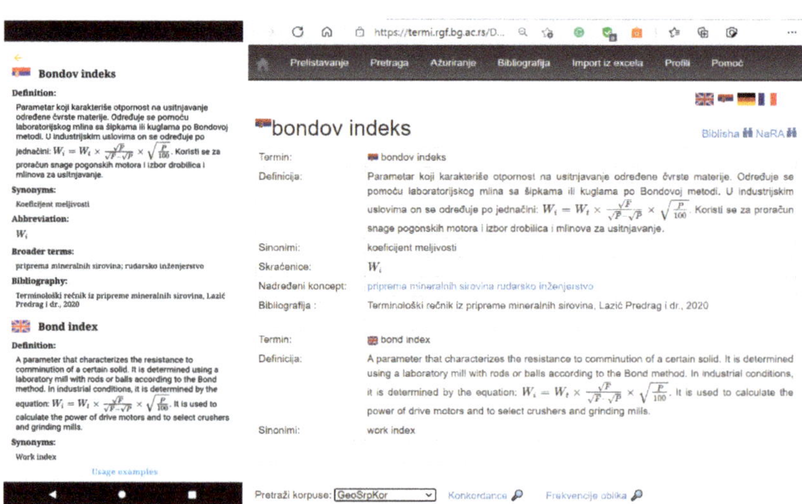

Figure 7. The mobile and Termi web application data entry preview for term entry.

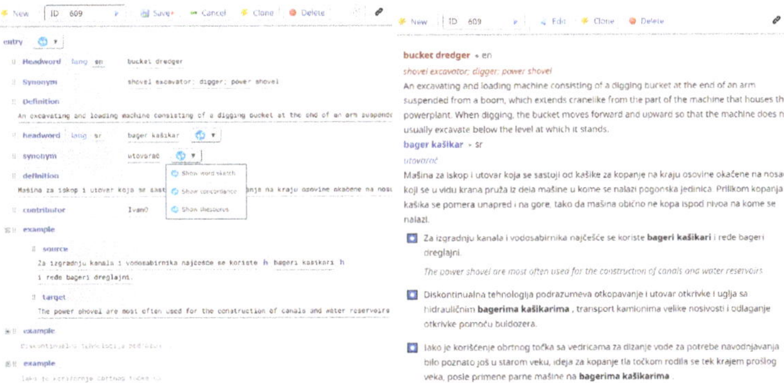

Figure 8. The Lexonomy data entry editing and preview for term entry.

5. Discussion

The presented approach to the development of terminology for the raw material domain, based on digitized and electronic dictionaries, terminological and domain corpora enables systematic development of terminology, complementing traditional terminological dictionaries with usage examples, and providing a comprehensive picture of the use of terms in various dictionaries, textbooks, professional and scientific literature. A terminology system that includes a relational terminology database, a SPARQL endpoint with linguistic linked open data, on the one hand, and a web and mobile application, on the other, provides a technological solution that enables data management, continuous updating, upgrading and expansion of available data, while various application forms (web and mobile) make the content more accessible to users.

Integration of terminology with the lexical database and morphological dictionaries, which enables support for a complex inflectional system, is important for all languages with rich morphology, such as Serbian. Integration with corpora, both standard and terminological, provides insight into the use of terms in modern language and in a specialized domain, enabling insight into individual examples, but also into the frequency of use of different syntactic structures, enabling research into collocations of individual terms.

The approach is demonstrated on the example of mining, but the same approach and developed software solutions can be used for other areas, which is certainly one of the further directions of activity. It should also be noted that the approach can be applied to other languages, depending on the available data and not on the language itself.

The vast amount of digitized resources, 22 dictionaries, monolingual corpus with 4 million words and bilingual with 12,657 aligned sentences, represent the basis for numerous other research activities, development of collocation dictionaries, creation of possibly printed dictionaries of different volumes (including pocket and encyclopedic ones). Such a system will make it easier for students to translate from English with the use of correct terms in Serbian, but also when writing articles and translating into English for academic purposes.

Since the presented approach used a combination of reuse of data, automatic extraction and manual post-editing, a comparison of those aspects with some similar solutions follows.

When it comes to the reuse of data, we followed the idea of the Sõnaveeb language portal of the Institute of the Estonian Language [10], which contains data from 70 dictionaries and termbases, comprising a total of 200,000 Estonian headwords with many new types of lexicographic information: collocations, etymology, multi-word expressions, and so forth. The number of lexicons in our case is much smaller, but at the moment we are focused on the mining domain and related terminology. Also, our system does not include etymology, but we plan to introduce it in the future. There is a difference in the software solution for mobile users, as Institute of the Estonian Language decided to produce a responsive web page that adapts to different devices by automatically adapting to the screen, whether it is a desktop, laptop, tablet or smartphone, while we produce a mobile android application akin to Oxford Dictionary or Merriam-Webster. Finally, the difference related to corpus use is that our system has direct connection with corpora, both domain and general language, which allows users to retrieve concordances, collocations defined by syntactic patterns and graphical frequency presentations. The Sõnaveeb project is a result of several projects in a longer period, developed by a much bigger team, but we are following their ideas to continually improve our system.

An Integrated Approach to Biomedical Term Identification Systems [11] combines several sources of information and knowledge bases to provide biomedical term identification systems with modular architecture, which includes medical term identification, retrieval of literature and ontology browsing by applying several NLP technologies. The similarity with our system is in combining several terminological and lexical resources, as well as the use of various NLP techniques, while the difference is that their system generates a conceptual graph that semantically relates all the terms found in the text, which would be our plan for future research. On the other side, our system is building a new resource that integrates a number of digitized and electronic resources.

The corpus-based approach for extracting domain-oriented and technical words applied to improve the efficiency of corpus analysis in COVID-19 big textual data [7] is based on elimination of function words and meaningless words. This, widely accepted, approach for information retrieval is not so successful for knowledge extraction, lexicographic and terminological purposes, so we are relying on a combination of syntactic patterns [34,42,68] and statistical association measures for domain terms: log-likelihood [69], c-value/nc-value [70], because such hybrid systems have proved to yield the best solutions [71].

Besides monolingual term extraction, we also followed a different approach when it comes to bilingual term extraction [72,73]. We first perform monolingual extraction of domain-specific terms, using available terminology extractors, and then, given a source term and a parallel sentence pair in which it appears, a set of possible translations are obtained. There are different options: to use automatic translation, trained on the same corpus using GIZA++ [40,43], to apply a word aligner [72], or to use log-likelihood comparison and phrase-based statistical machine translation models as in TermFinder [73]. We rely on previous research [27,39,40] that proved successful for bilingual term extraction in other domains, where one language is Serbian.

The Sketch-engine [9] has different types of extraction implemented, for various languages, starting with keyword extraction, word sketches, usage examples, and thesaurus, but it is not fully adapted for Serbian, and its results are far less successful than those obtained in our research [40,68]. Sketch Engine offers tools to significantly speed up the process of dictionary building, especially the "OneClick Dictionary" process, which consists of generating a headword list, providing part-of-speech labels, usage labels, generating candidates for example sentences, collocations, synonyms and thesaurus entries, definitions and/or translations [74]. The output is pushed into the Lexonomy dictionary writing system [66,67], from where lexicographers can communicate with the Sketch Engine during the post-editing phase, enabling browsing of concordances from a corpus and retrieval of selected examples directly into the interface form. The integration with corpus is a rare and very useful possibility, but Lexonomy lacks hierarchy browsing, mathematical formulae are not supported and search capabilities are limited.

6. Conclusions

The presented approach relies on the results of previous research in the field of NLP and terminology, but represents the first comprehensive solution for both building and using a terminology system that includes data, application and user interface layers covering different data and software technologies.

The automation of data publishing in the form of linked data, as one of the core pillars of the Semantic Web or the Web of Data, provides links between data sets that are understandable not only to humans, but also to machines, by sharing machine-readable interlinked data on the Web.

The next big challenge for the future is the automation of core lexicographic tasks related to semantics, such as finding definitions or identifying senses in two distinct processes: word-sense disambiguation (attributing the correct sense from a predefined set of senses) and word-sense induction (clustering of senses based on word context). Also, integration of results into linked open data especially word embeddings, collocation and similarities.

In future research we will incorporate synonyms for lexical sememe (smallest semantic unit for describing real-world concepts) prediction using an attention-based model [75], which scores candidate sememes from synonyms, by combining distances of words in the embedding vector space, and derives an attention-based strategy to dynamically balance two kinds of knowledge from a synonymous word set and word embedding vector.

Author Contributions: Conceptualization, O.K. and R.S.; Data curation, A.T. and L.K.; Formal analysis, R.S. and M.Š.; Investigation, O.K.; Methodology, O.K., R.S. and I.B., Validation, A.T. and L.K.; Software, O.K., M.Š. and I.B.; Writing—all authors. All authors have read and agreed to the published version of the manuscript.

Funding: This research was funded by Finnish Work Environment Fund and Ministry of Education, Science and Technological Development Republic of Serbia within European Science Program SAFERA (European Research on Industrial Safety towards Smart and Sustainable Growth) grant SafePotential, for period 2019–2020. Access to SketchEngine and Lexonomy is provided by the ELEXIS project funded by the European Union's Horizon 2020 research and innovation programme under grant number 731015. Linked data development is supported by the COST Action CA18209-NexusLinguarum "European network for Web-centred linguistic data science".

Institutional Review Board Statement: Not applicable.

Informed Consent Statement: Not applicable.

Data Availability Statement: The data presented in this study are freely available for the search on site: https://termi.rgf.bg.ac.rs/ (accessed on 15 March 2021) and on request from the corresponding author.

Acknowledgments: The authors thank Ivan Obradović for proofreading and constructive comments, Cvetana Krstev for use of electronic dictionary of Serbian, Petar Popović for corpus management and Branislava Šandrih for feature extraction from usage examples.

Conflicts of Interest: The authors declare no conflict of interest. The funders had no role in the design of the study; in the collection, analyses, or interpretation of data; in the writing of the manuscript, or in the decision to publish the results.

References

1. Prćić, T. *Ka Savremenim Srpskim rečNicima, Prvo, Elektronsko, Izdanje [Towards Modern Serbian Dictionaries, The First Digital Edition]*; Filozofski fakultet: Novi Sad, Srbija, 2018; p. 222.
2. ELEXIS—European Lexicographic Infrastructure. Available online: https://elex.is/ (accessed on 12 February 2020).
3. Smolka, E.; Schulte im Walde, S. *The Role of Constituents in Multiword Expressions: An Interdisciplinary, Cross-Lingual Perspective*; Language Science Press: Berlin, Germany, 2020; Volume 4. [CrossRef]
4. Kallas, J.; Koeva, S.; Langemets, M.; Tiberius, C.; Kosem, I. Lexicographic Practices in Europe: Results of the ELEXIS Survey on User Needs. In *Electronic Lexicography in the 21st Century, Proceedings of the eLex 2019 Conference, Sintra, Portugal, 1–3 October 2019*; Kosem, T., Kuhn, Z., Correia, M., Ferreria, J.P., Jansen, M., Pereira, I., Kallas, J., Jakubíček, M., Krek, S., Tiberius, C., Eds.; Lexical Computing: Brno, Czechia, 2019; pp. 1–3.
5. Krek, S. Natural Language Processing and Automatic Knowledge Extraction for Lexicography. *Int. J. Lexicogr.* **2019**, *32*, 115–118. [CrossRef]
6. Van der Merwe, M.F.; Horn, K. Mobile Concepts in a Mobile Environment: Historical Terms in LSP Lexicography. *Yesterday Today* **2018**, 17–34. [CrossRef]
7. Chen, L.C.; Chang, K.H.; Chung, H.Y. A Novel Statistic-Based Corpus Machine Processing Approach to Refine a Big Textual Data: An ESP Case of COVID-19 News Reports. *Appl. Sci.* **2020**, *10*, 5505. [CrossRef]

8. Granger, S. Electronic lexicography: From challenge to opportunity. In *Electronic Lexicography, Chapter Introduction*; Oxford University Press: Oxford, UK, 2012; pp. 1–15. [CrossRef]
9. Jakubiček, M.; Měchura, M.; Kovář, V.; Rychlý, P. Practical Post-Editing Lexicography with Lexonomy and Sketch Engine. In Proceedings of the XVIII EURALEX International Congress: Lexicography in Global Contexts, Book of Abstracts, Ljubljana, Slovenia, 17–21 July 2018; Cibej, J., Gorjanc, V., Kosem, I., Krek, S., Eds.; Ljubljana University Press: Ljubljana, Slovenia, 2018; pp. 65–67.
10. Koppel, K.; Tavast, A.; Langemets, M.; Kallas, J. Aggregating dictionaries into the language portal Sõnaveeb: Issues with and without a solution. Electronic Lexicography in the 21st Century: Smart Lexicography. In Proceedings of the eLex 2019 Conference, Sintra, Portugal, 1–3 October 2019; pp. 1–3.
11. Lopez-Ubeda, P.; Díaz-Galiano, M.C.; Montejo-Raez, A.; Martin-Valdivia, M.T.; Ureña-López, L.A. An integrated approach to biomedical term identification systems. *Appl. Sci.* **2020**, *10*, 1726. [CrossRef]
12. Graham, R.L. *Dictionary of Mining, Mineral & Related Terms*; US Bureau of Mines: Washington, DC, USA, 1996.
13. Dictionary of Mining, Mineral and Related Terms (an EduMine Tool). Available online: http://xmlwords.infomine.com/xmlwords.htm (accessed on 10 January 2020).
14. Nešić, G. *Rudarski rečNik: Srpsko-Hrvatski: English: Français: Deutsch: Russkij [Mining Dictionary: Serbo-Croatian: English: French: German: Russian]*; Rudarski Institut [Mining Institute]: Beograd, Srbija, 1970; p. 1291.
15. Lazić, S. *Englesko-Hrvatsko-Srpski Naftni Rječnik [English Croatian-Serbian Petroleum Dictionary]*; Poslovno udruženje Nafta: Zagreb, Hrvatska, 1976; p. 444.
16. Perić, M. *Englesko-Hrvatski Enciklopedijski rječNik Istraživanja i Proizvodnje Nafte i Plina [English-Croatian Encyclopedic Dictionary of Petroleum Exploration & Production]*; INA Industrija nafte d.d. Sektor korporativnih komunikacija: Zagreb, Hrvatska, 2007; p. 1038.
17. Tomanec, R.; Miljanović, I. *Mali Terminološki rečNik Pripreme Mineralnih Sirovina: Englesko-Srpski i Srpsko-Engleski [A small Dictionay of Mineral Preparation: English-Serbian and Serbian-English]*; Rudarsko-geološki fakultet [Faculty of Mining and Geology]: Beograd Srbija, 2002.
18. Lazić, P. *Terminološki Rečnik iz Pripreme Mineralnih Sirovina [The Terminological Dictionary of the Mineral Processing]*; Univerzitet u Beograd, Rudarsko-geološki fakultet [University of Belgradem Faculty of Mining and Geology]: Beograd, Srbija, 2020; Draft Version.
19. Stanković, R.M.; Obradović, I.; Kitanović, O.; Kolonja, L. Towards a mining equipment ontology. In Proceedings of the 12th International Conference 'Research and Development in Mechanical Industry' (RaDMI 2012), Vrnjačka Banja, Serbia, 14–17 September 2012; SaTCIP (Scientific and Technical Center for Intellectual Property) Ltd., Serbia: Vrnjačka Banja, Serbia, 2012; pp. 108–118.
20. Obradović, I.; Stanković, R.; Prodanović, J.; Kitanović, O. A TEL platform blending academic and entrepreneurial knowledge. In Proceedings of the Fourth International Conference on e-Learning (eLearning-2013), Manama, Bahrain, 7–9 May 2013; Belgrade Metropolitan University Belgrade: Belgrade, Serbia, 2013; pp. 65–70.
21. Stanković, R.; Obradović, I.; Utvić, M. Developing Termbases for Expert Terminology under the TBX Standard. In *Natural Language Processing for Serbian-Resources and Applications, Proceedings of the 35th Anniversary of Computational Linguistics in Serbia, Belgrade, Serbia, 12 November 2013*; Pavlović Lažetić, G., Vitas, D., Krstev, C., Eds.; University of Belgrade: Belgrade, Serbia, 2013.
22. Kolonja, L.; Stanković, R.; Obradović, I.; Kitanović, O.; Cvjetić, A. Development of terminological resources for expert knowledge: A case study in mining. *Knowl. Manag. Res. Pract.* **2016**, *14*, 445–456. [CrossRef]
23. Beko, L.; Obradović, I.; Stanković, R. Developing Students' Mining and Geology Vocabulary Through Flashcards and L1 in the CLIL Classroom. In Proceedings of the Second International Conference on Teaching English for Specific Purposes and New Language Learning Technologies, Niš, Serbia, 22–24 May 2015; Faculty of Electronic Engineering, University of Niš: Niš, Serbia, 2015.
24. Termi—Terminological Web Application. Available online: https://termi.rgf.bg.ac.rs/ (accessed on 12 February 2020).
25. GeolISS—Geološki Informacioni Sistem Srbije [Geological Information System of Serbia]—GeoliSSTerm. Available online: http://geoliss.mre.gov.rs/recnik/ (accessed on 12 February 2020).
26. Rudarska Terminologija i Nomenklatura [Mining Terminology and Nomenclature]. Available online: http://rudonto.rgf.bg.ac.rs/ (accessed on 12 February 2020).
27. Andonovski, J.; šandrih, B.; Kitanović, O. Bilingual lexical extraction based on word alignment for improving corpus search. *Electron. Libr.* **2019**, *37*, 722–739. [CrossRef]
28. Radojičić, M.; Obradović, I.; Stanković, R.; Utvić, M.; Kaplar, S. A Mathematical Learning Environment Based on Serbian Language Resources. In Proceedings of the 7th International Scientific Conference Technics and Informatics in Education, Čačak, Serbia, 25–27 May 2018; Faculty of Technical Sciences: Čačak, Serbia, 2018; pp. 248–254.
29. Stanković, R.; Krstev, C.; Lazić, B.; Vorkapić, D. A bilingual digital library for academic and entrepreneurial knowledge management. In Proceeding of the 10th International Forum on Knowledge Asset Dynamics-IFKAD, Bari, Italy 10–12 June 2015; pp. 1764–1777.
30. Stanković, R.; Krstev, C.; Vitas, D.; Vulović, N.; Kitanović, O. Keyword-based search on bilingual digital libraries. In *Semanitic Keyword-Based Search on Structured Data Sources*; Calì, A., Gorgan, D., Ugarte, M., Eds.; Springer: Berlin/Heidelberg, Germany, 2016; pp. 112–123.

31. Tomašević, A.; Stanković, R.; Utvić, M.; Obradović, I.; Kolonja, B. Managing mining project documentation using human language technology. *Electron. Libr.* **2018**, *36*, 993–1009. [CrossRef]
32. Stanković, R.; Krstev, C.; Lazić, B.; Škorić, M. Electronic Dictionaries—from File System to lemon Based Lexical Database. In Proceedings of the Eleventh International Conference on Language Resources and Evaluation—W23 6th Workshop on Linked Data in Linguistics: Towards Linguistic Data Science (LDL-2018), LREC 2018, Paris, France, 12 May 2018; McCrae, J.P., Chiarcos, C., Declerck, T., Gracia, J., Klimek, B., Eds.; European Language Resources Association (ELRA): Paris, France, 2018; pp. 18–23.
33. Krstev, C. *Processing of Serbian—Automata, Texts and Electronic Dictionaries*; Faculty of Philology of the University of Belgrade: Belgrade, Serbia, 2008.
34. Krstev, C.; Stanković, R.; Obradović, I.; Lazić, B. Terminology Acquisition and Description Using Lexical Resources and Local Grammars. In Proceedings of the 11th International Conference on Terminology and Artificial Intelligence, Granada, Spain, 4–6 November 2015; Volume 1495, pp. 81–89.
35. Stankovic, R.; Šandrih, B.; Krstev, C.; Utvić, M.; Škorić, M. Machine Learning and Deep Neural Network-Based Lemmatization and Morphosyntactic Tagging for Serbian. In Proceedings of The 12th Language Resources and Evaluation Conference, Marseille, France, 11–16 May 2020; European Language Resources Association: Marseille, France, 2020; pp. 3954–3962.
36. Schmid, H. Improvements in Part-of-Speech Tagging with an Application to German. In *Natural Language Processing Using Very Large Corpora*; Armstrong, S., Church, K., Isabelle, P., Manzi, S., Tzoukermann, E., Yarowsky, D., Eds.; Springer Netherlands: Dordrecht, The Netherland, 1999; pp. 13–25. [CrossRef]
37. Schmid, H. Probabilistic Part-of-Speech Tagging Using Decision Trees. In *New Methods in Language Processing*; Jones, D.B., Somers, H.L. Eds.; Routledge: London, UK, 2013; pp. 154–164.
38. Bilte—Bilingual Terminology Extraction. Available online: http://bilte.jerteh.rs/ (accessed on 12 February 2020).
39. Krstev, C.; Šandrih, B.; Stanković, R.; Mladenović, M. Using English baits to catch Serbian multi-word terminology. In Proceedings of the Eleventh International Conference on Language Resources and Evaluation (LREC 2018), Miyazaki, Japan, 7–12 May 2018; European Language Resources Association (ELRA): Miyazaki, Japan, 2018.
40. šandrih, B.; Krstev, C.; Stanković, R. Two approaches to compilation of bilingual multi-word terminology lists from lexical resources. *Nat. Lang. Eng.* **2020**, *26*, 455–479. [CrossRef]
41. Spasić, I.; Greenwood, M.; Preece, A.; Francis, N.; Elwyn, G. FlexiTerm: A flexible term recognition method. *J. Biomed. Semant.* **2013**, *4*, 27. [CrossRef] [PubMed]
42. Cram, D.; Daille, B. Terminology extraction with term variant detection. In Proceedings of ACL-2016 System Demonstrations, Berlin, Germany, 7–12 August 2016; Association for Computational Linguistics: Berlin, Germany, 2016; pp. 13–18. [CrossRef]
43. Och, F.J.; Ney, H. A systematic comparison of various statistical alignment models. *Comput. Linguist.* **2003**, *29*, 19–51. [CrossRef]
44. Moses–Statistical Machine Translation System. Available online: http://www.statmt.org/moses/ (accessed on 12 February 2020).
45. Mihaljević, M. Definicija naziva u terminološkim rječnicima [Names definition in terminological dictionaries]. *Raspr. čAsopis Instituta Hrvat. Jez. Jezikoslovlje Discuss. J. Inst. Croat. Lang. Linguist.* **1993**, *19*, 259–271.
46. Chiarcos, C.; Fäth, C.; Ionov, M. The ACoLi dictionary graph. In Proceedings of the 12th Conference on Language Resources and Evaluation (LREC 2020), Marseille, France, 11–16 May 2020; European Language Resources Association: Marseille, France, 2020; pp. 3281–3290.
47. The OntoLex Lemon Lexicography Module. Available online: https://www.w3.org/2019/09/lexicog/ (accessed on 12 February 2020).
48. Bosque-Gil, J.; Gracia, J.; Montiel-Ponsoda, E. Towards a Module for Lexicography in OntoLex. In Proceedings of the LDK workshops: OntoLex, TIAD and Challenges for Wordnets at 1st Language Data and Knowledge conference (LDK 2017), Galway, Ireland, 18 June 2017; Volume 1899, pp. 74–84.
49. Lexicon Model for Ontologies. Available online: https://www.w3.org/2016/05/ontolex/ (accessed on 12 February 2020).
50. McCrae, J.; Aguado-de Cea, G.; Buitelaar, P.; Cimiano, P.; Declerck, T.; Gómez-Pérez, A.; Gracia, J.; Hollink, L.; Montiel-Ponsoda, E.; Spohr, D.; et al. Interchanging lexical resources on the Semantic Web. *Lang. Resour. Eval.* **2012**, *46*, 701–719. [CrossRef]
51. NexusLinguarum COST Action. Available online: https://nexuslinguarum.eu/ (accessed on 12 February 2020).
52. RDF 1.1 Turtle. Available online: https://www.w3.org/TR/turtle/ (accessed on 12 February 2020).
53. Chiarcos, C.; Ionov, M.; de Does, J.; Depuydt, K.; Khan, F.; Stolk, S.; Declerck, T.; McCrae, J.P. Modelling Frequency and Attestations for OntoLex-Lemon. In Proceedings of the 2020 Globalex Workshop on Linked Lexicography, Marseille, France, 12 May 2020; pp. 1–9.
54. Stanković, R.; Šandrih, B.; Stijović, R.; Krstev, C.; Vitas, D.; Marković, A. SASA Dictionary as the Gold Standard for Good Dictionary Examples for Serbian. In *Electronic lexicography in the 21st Century, Proceedings of the eLex 2019 Conference, Sintra, Portugal, 1–3 October 2019*; Lexical Computing CZ, s.r.o.: Brno, Czech Republic, 2019; pp. 248–269.
55. Sketch Engine API. Available online: https://www.sketchengine.eu/documentation/api-documentation/ (accessed on 12 February 2020).
56. Kilgarriff, A.; Baisa, V.; Bušta, J.; Jakubíček, M.; Kovář, V.; Michelfeit, J.; Rychlý, P.; Suchomel, V. The Sketch Engine: Ten years on. *Lexicography* **2014**, *1*, 7–36. [CrossRef]
57. Sketchengine API for OntoLex FrAC Module. Available online: https://github.com/ontolex/frequency-attestation-corpus-information/blob/master/samples/sketch-engine/Sketch-Api-Frac.ipynb (accessed on 12 February 2020).

58. Statistics Used in the Sketch Engine. Available online: https://www.sketchengine.eu/wp-content/uploads/ske-statistics.pdf (accessed on 12 February 2020).
59. VocBench—Collaborative Development Platform for Managing Ontologies, Thesauri, Ontolex-Lemon Lexicons and Generic RDF Datasets. Available online: http://vocbench.uniroma2.it (accessed on 12 February 2020).
60. Stellato, A.; Fiorelli, M.; Turbati, A.; Lorenzetti, T.; van Gemert, W.; Dechandon, D.; Laaboudi-Spoiden, C.; Gerencsér, A.; Waniart, A.; Costetchi, E.; et al. VocBench 3—A collaborative Semantic Web editor for ontologies, thesauri and lexicons. *Semant. Web* **2020**, *11*, 855–881. [CrossRef]
61. Leximirka. Available online: https://leximirka.jerteh.rs (accessed on 12 February 2020).
62. MathJax—Open Source, JavaScript Display Engine for Mathematics that Works in all Browsers. Available online: https://www.mathjax.org/ (accessed on 12 February 2020).
63. Cervone, D. MathJax: A Platform for Mathematics on the Web. *Not. Am. Math. Soc.* **2012**, *59*, 312–316. [CrossRef]
64. KaTeX—Javascript Library for TeX Math Rendering on the Web. Available online: https://github.com/KaTeX/KaTeX (accessed on 12 February 2020).
65. KaTeX View—A Library that Uses Khan Academy KaTeX for TeX Math Rendering. Available online: https://github.com/judemanutd/KaTeXView (accessed on 12 February 2020).
66. Lexonomy. Available online: https://www.lexonomy.eu/ (accessed on 12 February 2020).
67. Měchura, M.B. Introducing Lexonomy: An open-source dictionary writing and publishing system. Electronic Lexicography in the 21st Century: Lexicography from Scratch. In Proceedings of the eLex 2017 Conference, Leiden, The Netherlands, 19–21 September 2017; pp. 19–21.
68. Stanković, R.; Krstev, C.; Obradović, I.; Lazić, B.; Trtovac, A. Rule-based automatic multi-word term extraction and lemmatization. In Proceedings of the Tenth International Conference on Language Resources and Evaluation (LREC'16), Portorož, Slovenia, 23–28 May 2016; pp. 507–514.
69. Gelbukh, A.; Sidorov, G.; Lavin-Villa, E.; Chanona-Hernandez, L. Automatic term extraction using log-likelihood based comparison with general reference corpus. In *International Conference on Application of Natural Language to Information Systems*; Springer: Berlin/Heidelberg, Germany, 2010; pp. 248–255.
70. Frantzi, K.; Ananiadou, S.; Mima, H. Automatic recognition of multi-word terms:. the c-value/nc-value method. *Int. J. Digit. Libr.* **2000**, *3*, 115–130. [CrossRef]
71. Pazienza, M.T.; Pennacchiotti, M.; Zanzotto, F.M. Terminology extraction: An analysis of linguistic and statistical approaches. In *Knowledge Mining*; Springer: Berlin/Heidelberg, Germany, 2005; pp. 255–279.
72. Arcan, M.; Turchi, M.; Tonelli, S.; Buitelaar, P. Leveraging bilingual terminology to improve machine translation in a CAT environment. *Nat. Lang. Eng.* **2017**, *23*, 763–788. [CrossRef]
73. Haque, R.; Penkale, S.; Way, A. TermFinder: Log-likelihood comparison and phrase-based statistical machine translation models for bilingual terminology extraction. *Lang. Resour. Eval.* **2018**, *52*, 365–400. [CrossRef]
74. Sketch Engine—Revolutionize the Dictionary-Building Process. Available online: https://www.sketchengine.eu/user-guide/lexicographers/ (accessed on 15 February 2021).
75. Kang, X.; Li, B.; Yao, H.; Liang, Q.; Li, S.; Gong, J.; Li, X. Incorporating Synonym for Lexical Sememe Prediction: An Attention-Based Model. *Appl. Sci.* **2020**, *10*, 5996. [CrossRef]

Article

Neuro-Fuzzy Transformation with Minimize Entropy Principle to Create New Features for Particulate Matter Prediction

Krittakom Srijiranon [1] and Narissara Eiamkanitchat [2,*]

1 Department of Computer Engineering, Faculty of Engineering, Graduate School, Chiang Mai University, Chiang Mai 50200, Thailand; krittakom@cs.tu.ac.th
2 Department of Computer Engineering, Faculty of Engineering, Chiang Mai University, Chiang Mai 50200, Thailand
* Correspondence: narisara@eng.cmu.ac.th

Abstract: Air pollution is a major global issue. In Thailand, this issue continues to increase every year, similar to other countries, especially during the dry season in the northern region. In this period, particulate matter with aerodynamic diameters smaller than 10 and 2.5 micrometers, known as PM_{10} and $PM_{2.5}$, are important pollutants, most of which exceed the national standard levels, the so-called Thailand air quality index (T-AQI). Therefore, this study created a prediction model to classify T-AQI calculated from both types of PM. The neuro-fuzzy model with a minimum entropy principle model is proposed to transform the original data into new informative features. The processes in this model are able to discover appropriate separation points of the trapezoidal membership function by applying the minimum entropy principle. The membership value of the fuzzy section is then passed to the neural section to create a new data feature, the PM level, for each hour of the day. Finally, as an analytical process to obtain new knowledge, predictive models are created using new data features for better classification results. Various experiments were utilized to find an appropriate structure with high prediction accuracy. The results of the proposed model were favorable for predicting both types of PM up to three hours in advance. The proposed model can help people who are planning short-term outdoor activities.

Keywords: neuro-fuzzy; prediction model; air pollution; $PM_{2.5}$; PM_{10}

Citation: Srijiranon, K.; Eiamkanitchat, N. Neuro-Fuzzy Transformation with Minimize Entropy Principle to Create New Features for Particulate Matter Prediction. *Appl. Sci.* **2021**, *11*, 6590. https://doi.org/10.3390/app11146590

Academic Editor: Chuan-Ming Liu

Received: 24 June 2021
Accepted: 14 July 2021
Published: 17 July 2021

Publisher's Note: MDPI stays neutral with regard to jurisdictional claims in published maps and institutional affiliations.

Copyright: © 2021 by the authors. Licensee MDPI, Basel, Switzerland. This article is an open access article distributed under the terms and conditions of the Creative Commons Attribution (CC BY) license (https://creativecommons.org/licenses/by/4.0/).

1. Introduction

Air pollution is a major problem in public health that increases health impacts on both the cardiovascular and respiratory systems in humans [1]. There are many important air pollutants, including ground-level ozone (O_3), carbon monoxide (CO), nitrogen dioxide (NO_2), sulfur dioxide (SO_2), and particulate matter (PM), announced by the World Health Organization. However, PM exceeds both the national and international standards to the greatest extent compared with others [2]. The PM is a mixture of particles that it compounds and four types of components, namely, organic, inorganic, biological, and carbonaceous materials. The proportion of each component is different in each area [3]. Most of the PM is classified into two categories by size, which are based on health-related effects [4]. The size of PM affecting human health has an aerodynamic diameter of less than 10 μm, which can only be detected by an electron microscope. There are two major sizes of PM. First, coarse particulate matter called PM_{10} is PM with an aerodynamic diameter smaller than 10 μm. Another type is fine particulate matter called $PM_{2.5}$, which is PM with an aerodynamic diameter smaller than 2.5 μm [5,6]. However, there are other types of PM, such as PM_1 [7], which are excluded from this research due to air pollution standards.

Every year during the dry season, which begins in February, the upper northern region of Thailand is affected by air pollution problems from both types of PM and this problem ends when the rainy season begins [8]. Anthropogenic activities, both garbage and agricultural burning, are important sources that contribute to air pollution. After the harvest periods, farmers prepare their area for the next crop period by burning their crop residues [9]. Another source is wildfire from natural and human-made occurrences as this area is mostly covered with forests and mountains. Fire management is difficult due to many limitations, such as a lack of effective equipment [10]. There are many policies from the government to protect and prohibit burning. However, the air pollution problem does not seem to be improved.

In recent years, researchers have been focused on both processes and methods in data science to apply it in various applications, such as daily cattle health classification [11], tomography image analysis [12], and student dropout prediction [13]. For the air pollution problem, data science techniques can implement notification systems to alert people by predicting the upcoming air pollution level. Numerous research articles are interested in applying data science to the air pollution problem, especially both types of PM. They try to find both appropriate processes and methods to create prediction models with high model performance or computation time reduction for their desired output, such as PM concentrations, PM levels, or classes [14–16]. The popular models are multiple linear regression (MLR), autoregressive integrated moving average (ARIMA), and various types of artificial neural networks (ANNs).

MLR is a popular statistical model for comparing the model performance with the ANN, but the results showed that MLR is less effective than ANN [17–20]. ARIMA is a common model for time-series data. There are two interesting examples. The first example, a combination of MLR and ARIMA proposed by [21] was used to predict daily and monthly average PM_{10} concentrations in Delhi, India. The second example, using the output data from ARIMA as input features for MLR, was presented by [22]. In the article, ARIMA is used with the dataset, including seasonal features and the period of seasonal patterns, to predict hourly PM_{10} concentrations in Negeri Sembilan, Malaysia.

The ANN is the most popular model selected by many researchers as it outperforms other models. The presentation in [23] focusing on three cities of China proposed a combination of the rolling mechanism and gray model in the data preparation process and the ANN model was used in the prediction process. The result was a prediction of the daily average values of PM_{10} concentrations and PM_{10} classes, calculated from the China air quality index. A research article presented in [24] applied ANN to predict the highest daily PM_{10} concentration in Santiago, Chile. The rule-based classification is used from a combination of two models, ANN and K-nearest neighbor (K-NN), to improve model performance in the minor classes. There is another type of ANN, long short-term memory (LSTM), used by [25]. The research presented an appropriate LSTM structure to predict the daily average PM_{10} concentration in Seoul, South Korea.

Another type of ANN is a combination of ANN and fuzzy logic called neuro-fuzzy. Two research articles used neuro-fuzzy with the Tagaki-Sugeno system to predict daily average PM_{10} concentrations in Turkey. The output data from fuzzy logic was used as an input feature for ANN. In the fuzzy logic part, in [26], a bell-shaped membership function was selected, while in [27], the Gaussian membership function was selected. Moreover, neuro-fuzzy is more effective than the other classifiers, such as NN and the support vector machine, when using the standard datasets from UCI reported by [28–30]. Neuro-fuzzy was selected to be applied in various applications, such as the diffuse large B-cell lymphomas classification [31]. In addition, in [32], it was reported that the positions for changing slope in the fuzzy membership function are very important, so the minimum entropy principle (MEP) is applied to find these values.

This research proposes the neuro-fuzzy with the minimum entropy principle model for data transformation to create new informative features that are used to represent historical data. Moreover, the proposed transformation model can reduce concerns about bias in raw data. Finally, an ANN model is created for new informative features. The three- and five-class output data of this model are the hourly PM_{10} and $PM_{2.5}$ classes associated with the Thailand standard. The results of the model can be an application implemented to alert people and for short-term outdoor activity planning up to three hours in advance.

2. Materials and Methods

This section is divided into three subsections. The first subsection presents the details of the research areas and air quality standards in this research. The second subsection proposes the structure of the proposed model to create new informative features. The third subsection discusses the details of the prediction model to classify both types of PM.

2.1. Thailand Air Quality Index

The study area of this research is the upper northern part of Thailand due to the air pollution problem during summer every year. This area includes 8 provinces: Chiang Mai, Chiang Rai, Lampang, Lamphun, Mae Hongson, Nan, Phayao, Phrae, and Uttaradit. Only fixed-site data monitoring stations from the Pollution Control Department (PCD), Ministry of Natural Resources and Environment, Thailand, were selected to create a prediction model.

There are 14 fixed-site data monitoring stations in total; each province except Uttaradit has at least one station. The timing of raw data from these stations differs depending on the availability of recorded data from each location. However, the first date for most of the recordings is 1 January 2010 and the recording end date is 30 April 2018 (for additional details, see Appendix A). Considering the completeness of data, only one station per province was selected from all stations. Therefore, there were eight fixed-site data monitoring stations used in this research as follows:

- Yupparaj Wittayalai School, Chiang Mai (CHM-Yup);
- Natural Resources and Environment Office, Chiang Rai (CHR-Env);
- Lampang Meteorological Station, Lampang (LPA-Met);
- Provincial Administrative Stadium, Lamphun (LPH-Sta);
- Natural Resources and Environment Office, Mae Hongson (MHS-Env);
- Chaloem Phra Kiat Hospital, Nan (NAN-Hos);
- Knowledge Park, Nan (NAN-Hos);
- Phrae Meteorological Station, Phrae (PHA-Met).

Data from PCD were divided into two groups. The first group was meteorological, including wind speed (WS), wind direction (WD), relative humidity (RH), pressure (PR), rain (RA), temperature (TEMP), and solar radiation (SR). The other group was air pollution data, including PM_{10}, $PM_{2.5}$, ground-level ozone (O_3), carbon monoxide (CO), nitrogen monoxide (NO), nitrogen dioxide (NO_2), and sulfur dioxide (SO_2). Each station records different parameters (for additional details, see Appendix B). According to the investigation, it was found that 6 out of 8 stations with almost all parameters were collected, except $PM_{2.5}$, available in only two stations: CHM-Yup and NAN-Hos. In addition, the rain was excluded as an input feature in all data monitoring stations due to numerous zero values with more than 99% during the focus period of the experiment.

To report the levels of air pollution for people, an air quality index was used. Air pollution concentrations were divided into groups and represented by the color scheme. The number of groups and the range of concentrations in each group differed according to the law of each country. In Thailand, the PCD under the Thai government announced the Thai air quality index (T-AQI) [33] as a standard for classifying air quality. This index selects six air pollutions, namely, PM_{10}, $PM_{2.5}$, O_3, CO, NO_2, and SO_2. In T-AQI calculations, each air pollution was transformed to the T-AQI level by the corresponding equation, then the final T-AQI level reported to people was identified from the maximum value of T-AQI. Both types of PM often have the highest T-AQI levels compared to the other four air pollutions, so this research selected only two types of PM to create a prediction model. There are five groups of T-AQI; therefore, the meaning and ranges of each group were calculated from concentrations of both types of PM, as shown in Table 1.

Table 1. Definition and range of Thailand air quality index from PM_{10} and $PM_{2.5}$.

T-AQI Level	Concentrations (μg/m^3)		Meaning
	PM_{10}	$PM_{2.5}$	
1	0–50	0–25	Very good
2	51–80	26–37	Good
3	81–120	38–50	Good but unhealthy for Sensitive Groups
4	121–180	51–90	Unhealthy
5	>180	>90	Very unhealthy

2.2. *The Neuro-Fuzzy Transformation with Minimum Entropy Principle Model*

Data transformation is an important process in data science. This research proposes a neuro-fuzzy with minimum entropy principle (NFT-MEP) model for a novel data transformation. The flowchart of the proposed model is displayed in Figure 1, divided into four processes. First, the raw data from PCD used extract–transform–load (ETL) to create the dataset. This process used the scatter plot to divide input features into two groups. The first group is input features that can apply the fuzzy membership function (FMF) as Dataset-I and the second group is input features that cannot apply FMF as Dataset-II. Therefore, two datasets were created from ETL. Second, the minimum entropy principle was used to find the optimal positions of each FMF from Dataset-I and then membership values were created as Dataset-III. Third, both Dataset-II and Dataset-III were combined and then neural network (NN) models were utilized to output data. Finally, new informative features were generated from the output of the previous process. The additional details of each process are represented in each subsection.

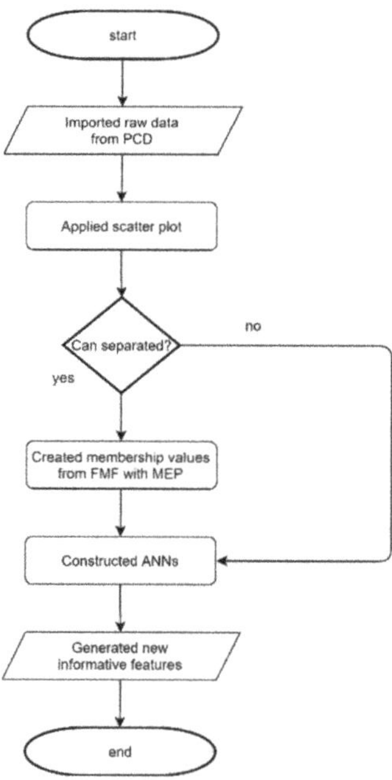

Figure 1. The flowchart of the neuro-fuzzy transformation with the minimum entropy principle model.

2.2.1. Extract–Transform–Load

The raw data from the PCD in each fixed-site data monitoring were received from different sensors, so all of them were extracted into a database and each database represents one station. Next, missing values were eliminated from the raw data. Each input feature was then considered to prepare for transformation. Scatter plots were utilized to input all features. They can determine the appropriate input features that can be transformed into membership values. The x-axis represents records of raw data and the y-axis represents the values of the input feature, while the colors of points represent the classes of PM. Considering that in each scatter plot, there is only one input feature that the distribution can separate from each color of the classes, it would be appropriate to use FMF to create membership values as Dataset-I. On the other hand, for an input feature that the distribution cannot separate from each color of the classes, the original value was used as Dataset-II. Finally, both Dataset-I and Dataset-II were loaded into the next process.

For example, the scatter plot of two input features from the LPA-Met station are shown, RH in Figure 2a and CO in Figure 2b, to filter out the appropriate features. This station contains approximately 16,000 records of raw data. The colors blue, red, and green, were used to represent three classes of the output data, Class 1, Class 2, and Class 3, respectively. As seen in Figure 2a, the scatter plot of the RH values and classes were difficult to separate from each other. On the other hand, the colors of the CO in Figure 2b were relatively separate. First, the blue color was mostly a CO value below 1. Second, the red color was mostly a CO value between 0.5 and 1.5. Finally, the green color was mostly a CO value above 1. Therefore, RH was loaded into Dataset-II, while CO was loaded into Dataset-I.

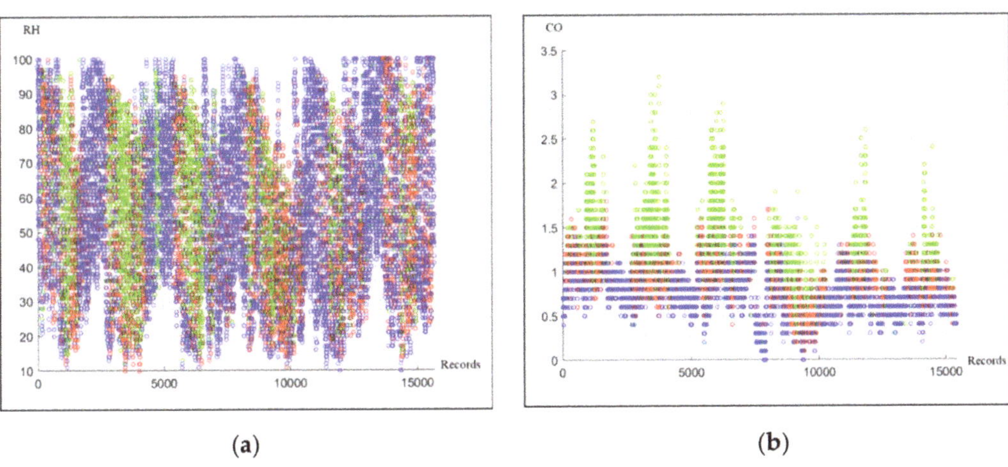

Figure 2. Scatter plot with three classes of PM_{10} in the LPA-Met station: (**a**) RH and (**b**) CO.

2.2.2. Fuzzy Membership Function with Minimum Entropy Principle

Fuzzy logic is based on uncertainty and an unsharp boundary that can be applied in some real-world applications. A difference between Boolean logic and fuzzy logic is that Boolean logic is a set of two values, completely true or 1 and completely false or 0. On the other hand, fuzzy logic is a fuzzy set including an infinite value between partial false or 0 and partial true or 1. Values in a fuzzy set called membership values are calculated by an FMF. This research selected trapezoidal functions as the FMF. Each input feature can include one or more FMFs and the number of FMFs of each input feature is two to five functions.

To find the optimal positions of the changing slope on the FMF, the minimum entropy principle (MEP) was used. This method finds the minimum value of entropy, which is an uncertainty of the data. The high entropy value means that there is a high probability that cannot divide data between classes. To find MEP, the threshold (x) in a range between X_1 and X_2 was calculated by Equations (1)–(3). This threshold divided data into two sides: the left side in $[X_1, x]$ as side p is calculated by Equation (1) and the right side in $[x, X_2]$ as side q is calculated by Equation (2). Then, x was gradually adjusted to the values between X_1

and X_2 to find the minimum entropy from Equation (3) and its value is the lowest entropy of data divided into two sides in ranges $[X_1, x]$ and $[x, X_2]$ [34].

$$S_p(x) = -\sum_{i=1}^{2} p_i(x) \ln p_i(x) \tag{1}$$

$$S_q(x) = -\sum_{i=1}^{2} q_i(x) \ln q_i(x) \tag{2}$$

$$S(x) = p(x)S_p(x) + q(x)S_q(x) \tag{3}$$

where $S(x)$ denotes the entropy value of x in range X_1 and X_2; $p(x)$ and $q(x)$ denote probabilities that all samples are in range $[X_1, x]$ and $[x, X_2]$, respectively; $p(x) + q(x) = 1$, $p_i(x)$ and $q_i(x)$ denote conditional probabilities that class i sample is in range $[X_1, x]$ and $[x, X_2]$, respectively.

After finding the minimum entropy as x_{min}, this value was used to determine the positions of the changing slope on the trapezoidal function by applying MEP again to find x_L and x_H. The x_L is a threshold with the minimum entropy in the range $[X_1, x_L]$ and $[x_L, x_{min}]$, while x_H is a threshold with the minimum entropy in range $[x_{min}, x_H]$ and $[x_H, X_2]$. Finally, x_L and x_H are separate points of the trapezoidal function. Next, the FMF was applied and each parameter has three to five new input features from the membership values.

Dataset-I from the ETL process applied FMF with MEP to create Dataset-III. For example, the CO in Figure 2b was applied to the MEP twice. The first MEP was used to divide between Class 1 and Class 2, while the second MEP was used to divide between Class 2 and Class 3. The first MEP results showed that x_L and x_H were 0.75 and 1.05 with the minimum entropy values 0.5165 and 0.6584, respectively. In addition, the second MEP results showed that x_L and x_H were 1.15 and 1.45 with the minimum entropy values 0.5595 and 0.4230, respectively.

This feature was divided into three FMFs. The membership values of each membership function were calculated from Equations (4)–(6) for low, medium, and high, respectively, where μ denotes the membership value and x denotes an input feature. In addition, Figure 3 shows a graph of three trapezoidal membership functions of the CO.

$$\mu_{Low}(x) = max\left(min\left(1, \frac{1.05 - x}{1.05 - 0.75}\right), 0\right) \tag{4}$$

$$\mu_{Medium}(x) = max\left(min\left(\frac{1.05 - x}{1.05 - 0.75}, 1, \frac{x - 1.15}{1.45 - 1.15}\right), 0\right) \tag{5}$$

$$\mu_{High}(x) = max\left(min\left(1, \frac{x - 1.15}{1.45 - 1.15}\right), 0\right) \tag{6}$$

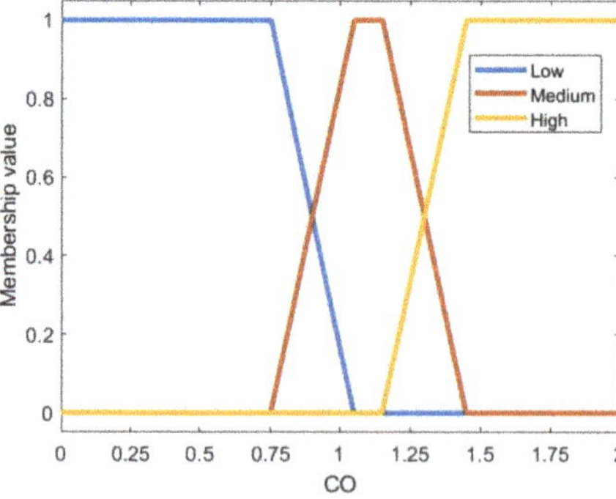

Figure 3. Fuzzy membership function for CO in the LPA-Met station.

As described earlier in the concept of selecting the appropriate input features, they were then selected for transformation by the fuzzy concept. Since raw data were checked at every station, the results of the selected input features were different for each station. Considering the selected input features, the meteorological data were inappropriate for transformation by the FMF. On the other hand, the air pollution data, especially CO, NO_x, and NO_2, were appropriate for transformation by the FMF. In addition, every station selected both types of PM to create the membership value.

2.2.3. Artificial Neural Networks

Artificial neural networks (ANNs) are a mathematical model that is imitated from the human nervous system. There are numerous neurons to process data. Neurons transfer data to one another. An advantage of ANN is that the parameters can be learned and modified from error. A popular structure of ANN is a combination of three types, including an input layer, hidden layers, and output layer. The input layer represents input features, while the output layer represents output classes. Each layer contains a group of neurons that receive information from the other neurons in the previous layer and send the information to the other neurons in the next layer [35].

The process of ANN is a combination of the set of input data and random weights plus the bias value. Next, the output value from the first process is transformed by a sigmoid transfer function. The output value after passing the transfer function is between 0 and 1. The ANN has self-adaptive learning, which adjusts all weight values from their error, called the backpropagation algorithm [36]. The stochastic gradient descent (SGD), among the popular weight optimization algorithms, was selected in this research to minimize the loss function, which is an error of the model. Finally, each weight value was updated by the chain rule of calculus.

This research enhanced the ANN structure proposed by [37]. In previous research, this model has been used to predict a daily average PM_{10} class where classes are defined according to the T-AQI. The structure of ANN is divided into two processes: the constructing an ANN model process and the decision process. In the first process, there are many ANN models and the number of models is equal to the number of classes. Each ANN model focuses on learning for each class, which includes an input layer, two hidden layers, and an output layer. For the input layer, Dataset-II and Dataset-III were combined and used as input features. The number of hidden neurons was fixed to six and three neurons in the first and second hidden layers, respectively. Finally, only one output neuron was utilized in the output layer. The initial parameters of ANN in every model were similar, including random weights for all neurons, a sigmoid transfer function for all layers, and a learning rate of -0.02. In the second process, the class in each record was identified by the outputs from the ANN models by Equation (7), where *Class* denotes the class of data and O_i denotes output data from ANN in model i. The ANN had the same number of classes. The value of output data of each model ranged from 0 to 1 due to the sigmoid transfer function. The maximum function determined the maximum value of the output data, then the index function was used to find the index of the maximum value. Finally, the class was identified by the index value.

$$Class = index(max(O_i)) \tag{7}$$

2.2.4. New Informative Features Generation

The original features of meteorological and air pollution data were applied to the processes described in Sections 2.2.1–2.2.3. The ANNs were then used to generate the historical situation of the PM level expressed by AQI relative to the desired class. Many research articles reported that historical data, both meteorological and air pollution data, affected the performance of the model [38–40], so this information was used to create new informative features. For the last process of the NF-MEP model, the output data from the ANN model at time $t-1$ to time $t-n$ were generated to predict the level of the PM at time t, where n denotes the number of hours prior.

An example of new informative features from NFT-MEP is shown in Figure 4. The table on the left of the figure illustrates the output data generated from the NFT-MEP model with five classes according to T-AQI. The first column shows the time in a 24-h cycle and the second column is the PM concentration (1–5). The table on the right of the figure illustrates an example of the six hours before the desired time dataset. The first column shows the desired prediction time and the next 6 columns are 1–6 h of concentration of the PM expressed in T-AQI. In predicting PM intensity at 9:00 a.m. on Day 1, the input characteristics generated from the NFT-MEP model were {4, 3, 2, 3, 3, 2}, representing the concentration data of 6 h prior, from 8.00 a.m. to 3.00 a.m. Four new datasets of the previous 6, 12, 18, and 24 h were created to determine the best historical period to use that provides the best prediction accuracy. The details and results of using these datasets are described in Section 3.1.

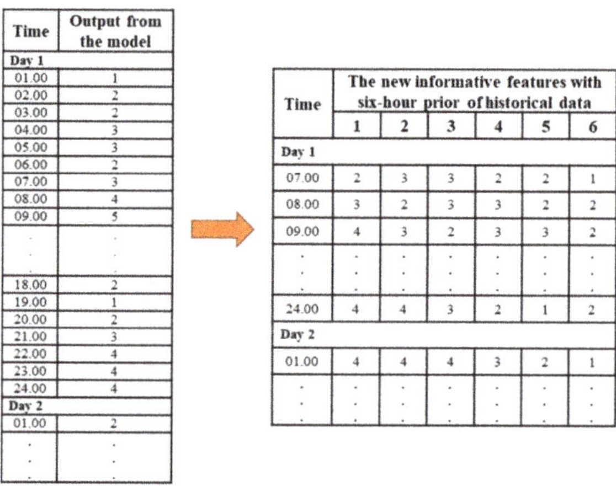

Figure 4. Example of new informative features with six-hour prior data created from the NFT-MEP model.

2.3. PM Prediction Model

The new informative features created from the NFT-MEP represent realistic data to improve prediction results. These features were used to construct a prediction model to classify the desired result. Another NN model was selected that was created from the new informative features. The structure of this model is similar to the structure of the NN model in the NFT-MEP model. In addition, the number of ANN models was three or five depending on the number of output classes. In general, the correct classification percentage is a popular statistical indicator to assess the performance of the model. However, the model in this research was an imbalanced classification problem, so two additional statistical indicators, F-score and Matthews correlation coefficient (MCC), were applied [41,42].

The output of the model is to predict the hourly T-AQI calculated from both types of PM. The hourly data can be used for short-term outdoor activity planning. The hourly PM_{10} and $PM_{2.5}$ concentrations were converted into classes according to the information in Table 2. This research selected two different types of output data, including three and five classes, during the experimental processes described in Section 3. For the three classes of output data, Class 1, which indicates "Good", was grouped according to the first two T-AQI levels. Second, Class 2, which indicates "Moderate (except for sensitive people)", was grouped according to T-AQI levels 3 and 4. Finally, Class 3, which indicates "Unhealthy", was the remaining level. The five classes of the output are the most detailed for implementation in real-world applications divided into five classes similar to the T-AQI level.

Table 2. The class assignments in the experiments were determined according to the standard Thailand PM concentrations.

<table>
<tr><th colspan="3">Three Classes of Output Data</th><th colspan="3">Five Classes of Output Data</th></tr>
<tr><th rowspan="2">Class Name</th><th colspan="2">Hourly Value</th><th rowspan="2">Class Name</th><th colspan="2">Hourly Value</th></tr>
<tr><th>PM_{10}</th><th>$PM_{2.5}$</th><th>PM_{10}</th><th>$PM_{2.5}$</th></tr>
<tr><td rowspan="2">Class 1</td><td rowspan="2">0–80</td><td rowspan="2">0–37</td><td>Class 1</td><td>0–50</td><td>0–25</td></tr>
<tr><td>Class 2</td><td>51–80</td><td>26–37</td></tr>
<tr><td>Class 2</td><td>81–120</td><td>38–50</td><td>Class 3</td><td>81–120</td><td>38–50</td></tr>
<tr><td rowspan="2">Class 3</td><td rowspan="2">>120</td><td rowspan="2">>50</td><td>Class 4</td><td>121–180</td><td>51–90</td></tr>
<tr><td>Class 5</td><td>>180</td><td>>90</td></tr>
</table>

3. Experimental Methods and Results

In this section, various experiments are presented to find the appropriate structure of the proposed model or to confirm model performance. The details of the experimental design consist of four subsections. The first three subsections are experiments to predict the class of PM one hour in advance. The first one found the best time interval for the new informative features. The second one was used to confirm that the new informative features created from FMF with MEP can increase the prediction performance. These experiments used four out of eight stations. The first two data monitoring stations were the CHM-Yup and NAN-Hos stations, due to the availability of the $PM_{2.5}$ data. The other two stations were the LPA-Met and PHY-Kno stations. The third subsection implemented the proposed model to all data monitoring stations and the overall model performances were reported. In addition, other popular prediction models in this problem were selected to compare the model performance with the proposed model. The last subsection was the reported model performance of the proposed model to predict an additional period of output data up to three hours in advance.

To obtain accurate prediction results, a specific data set for the dry season from 1 February to 31 May of each year, during which air pollution levels in Thailand are high, was the focus of this research. The dataset during the crisis of the last two years was defined as the testing data. The first set was raw data between 1 February 2018, and 30 April 2018. The second set was raw data between 1 February 2017, and 31 May 2017, while the remaining years were selected as the training data.

3.1. Experimental Method and Results for the New Informative Features with Different Number of Historical Data

This experiment aimed to determine an appropriate number of hours before the generation of the new informative features, as described in Section 2.2.4. The dataset of the five different time periods, 1, 6, 12, 18, and 24 h, was used in the experiments. Therefore, each dataset had a different number of features that varied from 1 to 12 depending on the number of hours prior. The experiments in this subsection used three classes that were defined per the T-AQI standard, as detailed in Table 2. The hourly PM_{10} class prediction was used in four stations, while the hourly $PM_{2.5}$ class prediction was used in two stations, due to the reason described earlier.

Table 3 shows the results of the class prediction of PM_{10} with the F-Score separated by class and the average overall and average accuracy of the two testing datasets. In addition, the PHY-Kno station had no experimental result from 24 h prior due to a lack of continuity data. The results shown in Table 3 in the last column show that the usage of 6 h usage had the highest F-score in three out of the four stations, CHM-Yup, NAN-Hos, and PHY-Kno stations. In the LPA-Met Station, there was no clear F-score result for any time period as with the other stations. In addition, 6 h prior had the highest average accuracy in every station.

Table 3. Model performance of the transformed dataset with different amounts of historical data to predict hourly PM_{10} with the three classes of output data.

Stations	Number of Hours Prior	Accuracy	F-Score			
			Class 1	Class 2	Class 3	Average
CHM-Yup	1	68.59%	0.7859	0.5457	0.3971	0.5762
	6	70.90%	0.7933	0.5810	0.4553	0.6099
	12	68.60%	0.7749	0.5554	0.4168	0.5824
	18	69.93%	0.7881	0.5688	0.4087	0.5885
	24	67.49%	0.7558	0.5571	0.4262	0.5797
NAN-Hos	1	80.19%	0.8888	0.4536	0.1657	0.5027
	6	83.86%	0.9092	0.5164	0.2613	0.5623
	12	83.02%	0.9037	0.5224	0.1497	0.5253
	18	81.82%	0.8957	0.5264	0.1344	0.5188
	24	82.15%	0.8856	0.5650	0.1652	0.5386
LPA-Met	1	80.09%	0.8765	0.6562	0.7063	0.7463
	6	81.70%	0.8748	0.7180	0.7604	0.7844
	12	78.47%	0.8690	0.7129	0.7702	0.7840
	18	81.67%	0.8742	0.7175	0.7683	0.7867
	24	80.65%	0.8648	0.7011	0.7614	0.7758
PHY-Kno	1	81.29%	0.9138	0.5405	0.5428	0.6657
	6	82.25%	0.8908	0.5973	0.6692	0.7191
	12	81.41%	0.8919	0.5729	0.5968	0.6872
	18	77.93%	0.8692	0.5288	0.6050	0.6677
	24	-	-	-	-	-

The same conditions were applied to experiments of the $PM_{2.5}$ datasets. Table 4 shows that the transformed dataset of 6 h prior had the highest average F-score in the CHM-Yup station, but this period had an inferior average F-score in the NAN-Hos station. The transformed dataset of 12 h prior had the highest average F-score in the NAN-Hos station. Considering the average accuracy, the transformed dataset of 6 h prior had the highest value in both stations. The results of the transformed dataset of 6 h prior showed that the average accuracy was 76.51% and 72.59% and the average F-score was 0.7194 and 0.5846 for CHM-Yup and NAN-Hos stations, respectively.

Table 4. Model performance of the transformed dataset with different amounts of historical data to predict hourly $PM_{2.5}$ with the three classes of output data.

Stations	Number of Hours Prior	Accuracy	F-Score			
			Class 1	Class 2	Class 3	Average
CHM-Yup	1	75.68%	0.7818	0.3728	0.8519	0.6689
	6	76.51%	0.7782	0.5068	0.8733	0.7194
	12	76.24%	0.7603	0.5152	0.8738	0.7165
	18	76.09%	0.7570	0.5127	0.8726	0.7141
	24	75.75%	0.7639	0.5041	0.8694	0.7125
NAN-Hos	1	66.06%	0.8156	0.1738	0.5570	0.5155
	6	72.59%	0.8619	0.2561	0.6356	0.5846
	12	72.05%	0.8468	0.2823	0.6656	0.5982
	18	71.24%	0.8360	0.2472	0.6370	0.5734
	24	64.04%	0.7861	0.3385	0.4664	0.5303

3.2. Experimental Method and Results of the Neuro-Fuzzy Transformation with and without MEP

The aim of the experiments in this section was to investigate whether adding FMF with MEP to the process and using those new informative features can improve prediction accuracy. The dataset of PM_{10} from the four stations was selected for this experiment. The 6 h prior dataset was built on the new features of NFT-MEP. Moreover, the structure from

Section 2.2, which excludes FMF with MEP as the neural network transformation (NT), was used in the experiment.

The comparison results of the NT model and the NFT-MEP model to predict hourly PM_{10} with three classes of output data are reported in Table 5, where all results were the averaged value between the two testing datasets. The results in Table 5 revealed that the NFT-MEP model had higher statistical indicators than the NT model in every station, which indicates that the neuro-fuzzy transformation gives better results than the one that is not used. Considering the performance of the model in each station, the NFT-MEP model had much better performance than the NT model in the CHM-Yup and NAN-Hos stations. On the other hand, this model slightly improved efficiency on the other two stations.

Table 5. Comparison result between the NT and NFT-MEP models to predict hourly PM_{10} with three classes of output data.

Stations	Model	Statistic Indicators	
		Accuracy	F-Score
CHM-Yup	NT	70.90%	0.6099
	NFT-MEP	81.99%	0.7012
NAN-Hos	NT	83.86%	0.5623
	NFT-MEP	90.83%	0.6253
LPA-Met	NT	81.70%	0.7844
	NFT-MEP	84.18%	0.7928
PHY-Kno	NT	82.25%	0.7191
	NFT-MEP	88.15%	0.7579

Next, the NFT-MEP model was used to predict hourly $PM_{2.5}$ with three classes of output data. The results found that the NFT-MEP model had higher statistical indicators than the NT model in every station similar to the PM_{10} model. The results of the NFT-MEP model were 81.45% and 85.29% for average accuracy and 0.7851 and 0.7824 for average F-score for CHM-Yup and NAN-Hos stations, respectively. The NFT-MEP model had a much-improved efficiency of the NT model, especially in the NAN-Hos station.

Finally, the results in this section showed that the NFT-MEP model had a higher model performance to predict hourly classes for both types of PM in every selected station than the NT model. Therefore, applying FMF with MEP to the NT model could improve the efficiency of the model. The average accuracy of the prediction model was more than 80% of both types of PM. In addition, the average F-scores of the prediction model was mostly greater than 0.7 for both types of PM, except the NAN-Hos station.

3.3. Comparison Results between the NFT-MEP Model and Other Popular Models

To verify the performance of the proposed NFT-MEP model, the other popular models in this problem were selected, including LSTM [15], ARIMA [12], and ARIMAX [34], for comparison. Every other model adjusted the structures to find appropriate parameters. The experimental design in this section differed from the previous section. Four additional stations, namely, CHR-Env, MHS-Env, LPH-Sta, and PHA-Met stations, were selected, so there were eight stations in this experiment. Moreover, the five classes of output data, for which the details are shown in Table 2, were selected to create a prediction model. Finally, each station was applied to four prediction models, namely, NFT-MEP, LSTM, ARIMA, and ARIMAX, and two different output data, including three and five classes. To compare model performance, three statistical indicators, namely, accuracy, F-score, and MCC, were used in this subsection.

The comparison results of the four models to predict hourly PM_{10} with three and five classes of output data are reported in Table 6. All results were an average value between two testing datasets from all stations. The results for the three classes of output data showed that the NFT-MEP model had the highest average accuracy with a value between

79.40% and 90.83%. In addition, the NFT-MEP model had the highest average F-score with a value between 0.6253 and 0.8183 and the highest average MCC between 0.5318 and 0.7395. The LSTM showed an inferior model performance to the NFT-MEP model, while the ARIMA and ARIMAX showed the lowest model performance mainly because they cannot classify Class 2 and Class 3. In addition, the results for the five classes of output data were similar to those of the three classes of output data. The results showed that the NFT-MEP model had the highest statistic indicators. The average accuracy of the NFT-MEP model was between 67.40% and 83.31%. In addition, the average F-score was between 0.5001 and 0.7255, and the average MCC was between 0.6778 and 0.4983. The LSTM had a higher model performance than the other two models.

Table 6. The comparison result of four prediction models to predict hourly PM_{10} with three and five classes of output data.

Types of Output Data	Model	Average Statistic Indicators		
		Accuracy	F-Score	MCC
Three-class	NFT-MEP	85.18%	0.7320	0.6361
	LSTM	80.98%	0.6478	0.4854
	ARIMAX	74.56%	0.5890	0.3860
	ARIMA	70.52%	0.4731	0.3197
Five-class	NFT-MEP	74.49%	0.6433	0.6035
	LSTM	62.71%	0.4385	0.3908
	ARIMAX	57.44%	0.3500	0.2848
	ARIMA	52.90%	0.2949	0.2666

The four models were used to predict hourly $PM_{2.5}$ with three and five classes of output data similar to PM_{10}, which are reported in Table 7. The results showed that the NFT-MEP model had the highest three statistic indicators compared to the three other models similar to the PM_{10} model. The average accuracy of the NFT-MEP model for the three classes of output data was between 81.45 and 85.28%. In addition, the average F-score was between 0.7824 and 0.7851, and the average MCC was between 0.6847 and 0.6920. The average accuracy of the NFT-MEP model for five classes of output data was between 73.76% and 76.16%. In addition, the average F-score was between 0.7229 and 0.7285 and the average MCC was between 0.6515 and 0.6632. For both types of output data, the LSTM had an inferior model performance and the other two models had the lowest model performance.

Table 7. Comparison result of four prediction models to predict hourly $PM_{2.5}$ with three and five classes of output data.

Types of Output Data	Model	Average Statistic Indicators		
		Accuracy	F-Score	MCC
Three-class	NFT-MEP	83.37%	0.7838	0.6883
	LSTM	77.57%	0.6879	0.5654
	ARIMAX	62.19%	0.5850	0.4859
	ARIMA	59.24%	0.5677	0.4587
Five-class	NFT-MEP	74.96%	0.7257	0.6573
	LSTM	62.12%	0.5989	0.4396
	ARIMAX	57.40%	0.4844	0.3570
	ARIMA	55.72%	0.4746	0.3021

As evidenced by the experimental results, the NFT-MEP model had the highest model performance. The LSTM had an inferior model performance, while ARIMA and ARIMAX had the lowest model performance. Based on the experimental results, it can be concluded that the NFT-MEP model outperformed both types of PM for prediction with two different amounts of output data when compared with the three other popular PM prediction models.

3.4. Implementation Results of the NFT-MEP Model to Predict Additional Periods of Output Data

From the previous experiment, the NFT-MEP model outperformed the other popular PM prediction models. However, this model predicts only one hour ahead of both types of PM. To implement the NFT-MEP model in real-world applications, information about PM one hour in advance was not sufficient for outdoor activity planning. This subsection implemented the NFT-MEP model to predict additional periods: two and three hours in advance. The implementation results are reported in Table 8. The results showed that as the length of the time periods increased, the model performance of the proposed model decreased for both types of PM and output data. However, the overall accuracy was more than 70 and 60% for three and five classes of output data, respectively. In addition, the F-score was more than 0.6 and MCC was approximately 0.5 for both types of PM.

Table 8. Implementation results to predict both types of PM with additional periods.

Types of PM	Types of Output Data	Periods of Output Data								
		One Hour in Advance			Two Hours in Advance			Three Hours in Advance		
		Accuracy	F-Score	MCC	Accuracy	F-Score	MCC	Accuracy	F-Score	MCC
PM_{10}	Three-class	85.18%	0.7320	0.6361	80.14%	0.6533	0.5712	76.25%	0.6001	0.5445
	Five-class	74.49%	0.6433	0.6035	67.48%	0.5465	0.4811	63.15%	0.4904	0.4233
$PM_{2.5}$	Three-class	83.37%	0.7838	0.6883	77.48%	0.7129	0.6249	73.10%	0.6645	0.5749
	Five-class	74.96%	0.7257	0.6573	65.44%	0.6233	0.5756	60.55%	0.5602	0.4984

4. Conclusions

This research proposed a novel approach of data transformation called neuro-fuzzy transformation with the minimum entropy principle. The proposed model was used to create new features for predicting classes of both types of PM. The raw data from eight fixed-site data monitoring stations were received from the PCD, Thailand, to create prediction models. Several experiments were conducted. The results showed that the new informative features of six hours prior were appropriate for the generation of historical data. In addition, the applied fuzzy membership function with the minimum entropy principle can improve model performance. It is evident from all experimental results that the proposed NFT-MEP model for data transformation outperformed in predicting both PM_{10} and $PM_{2.5}$ classes for all selected data monitoring stations.

Author Contributions: K.S. contributed to data acquisition, data analysis, model creation, and writing—original draft preparation; N.E. contributed to validation and writing—review and editing, supervision. Both authors have read and agreed to the published version of the manuscript.

Funding: This research received no external funding.

Institutional Review Board Statement: Not applicable.

Informed Consent Statement: Not applicable.

Data Availability Statement: The raw data of this work are available from the Pollution Control Department, Ministry of Natural Resources and Environment, Thailand, upon request.

Acknowledgments: This study was supported in part by The Graduate School, Chiang Mai University.

Conflicts of Interest: The authors declare that there is no conflict of interest regarding the publication of this paper.

Appendix A

The starting date of each station is as follows:

- The starting date of the CHM-Yup station was 21 May 2011, instead of 1 February 2010, as $PM_{2.5}$ data were available after 16 May 2011, and PM_{10} data were not available until 21 May 2011. In addition, PRES, RAIN, SR, and O_3 on the CHM-Yup station were not available after 21 August 2014, so they were excluded as input features.
- The starting dates of PHY-Kno and CHR-Env stations were 1 February 2011, as the first dates of the recorded data were June 2010, and July 2010, respectively.
- The starting dates of the NAN-Hos stations were 1 February 2016, as the first date of the recorded data was June 2015.
- The starting date of the LPH-Sta and MHS-Env station was 1 February 2010. In addition, the starting date of the PHA-Met station was 5 May 2010.
- The starting date of the LPA-Sta station was 1 February 2013, due to the first date that air pollution data were available. In addition, CO was excluded as an input feature as it was not available after 6 September 2017.

Appendix B

The input features list from sensors is shown in Table A1.

Table A1. List of parameters from eight stations divided by the type of data.

Station	Meteorological Data	Air Pollution
CHM-Yup	TP, RH, WS, WD	CO, SO_2, NO_X, NO, NO_2, $PM_{2.5}$
CHR-Env	TP, RH, PR, RA, WS, WD	CO, O_3
LPA-Met	TP, RH, PR, RA, SR, WS, WD	CO, SO_2, NO_X, NO, NO_2, O_3
LPH-Sta	TP, RH, PR, SR, WS, WD	SO_2, NO_X, NO, NO_2, O_3
MHS-Env	TP, RH, PR, RA, WS, WD	CO, O_3
NAN-Hos	TP, RH, PR, RA, SR, WS, WD	CO, SO_2, NO_X, NO, NO_2, O_3, $PM_{2.5}$
PHY-Kno	TP, RH, PR, RA, SR, WS, WD	CO, SO_2, NO_X, NO, NO_2, O_3
PHA-Met	TP, RH, PR, SR, WS, WD	CO, SO_2, NO_X, NO, NO_2, O_3

References

1. Bhat, T.H.; Jiawen, G.; Farzaneh, H. Air Pollution Health Risk Assessment (AP-HRA), Principles and Applications. *Int. J. Environ. Res. Public Health* **2021**, *18*, 1935. [CrossRef]
2. Ambient (Outdoor) Air Pollution. Available online: https://www.who.int/news-room/fact-sheets/detail/ambient-(outdoor)-air-quality-and-health (accessed on 1 May 2021).
3. Li, Z.; Wen, Q.; Zhang, R. Sources, health effects and control strategies of indoor fine particulate matter (PM2.5): A review. *Sci. Total Environ.* **2017**, *586*, 610–622. [CrossRef]
4. Gautam, S.; Patra, A.K.; Sahu, S.P.; Hitch, M. Particulate matter pollution in opencast coal mining areas: A threat to human health and environment. *Int. J. Min. Reclam. Environ.* **2016**, *32*, 75–92. [CrossRef]
5. WHO. Air Quality Guidelines for Particulate Matter, Ozone, Nitrogen Dioxide and Sulfur Dioxide. Available online: www.who.int/airpollution/publications/aqg2005/en/ (accessed on 7 October 2020).
6. Particulate Matter (PM) Pollution. Available online: https://Epa.gov/pm-pollution/particulate-matter-pm-basics (accessed on 7 November 2020).
7. Jakovljević, I.; Štrukil, Z.S.; Godec, R.; Bešlić, I.; Davila, S.; Lovrić, M.; Pehnec, G. Pollution Sources and Carcinogenic Risk of PAHs in PM1 Particle Fraction in an Urban Area. *Int. J. Environ. Res. Public Health* **2020**, *17*, 9587. [CrossRef]
8. Moran, J.; NaSuwan, C.; Poocharoen, O.-O. The haze problem in Northern Thailand and policies to combat it: A review. *Environ. Sci. Policy* **2019**, *97*. [CrossRef]
9. Punsompong, P.; Chantara, S. Identification of potential sources of PM10 pollution from biomass burning in northern Thailand using statistical analysis of trajectories. *Atmos. Pollut. Res.* **2018**, *9*, 1038–1051. [CrossRef]
10. Homhuan, S.; Humhong, C. The development of forest fire monitoring and warning system for agroforestry areas in Uttaradit Province, Thailand. In *IOP Conference Series: Earth and Environmental Science*; IOP Publishing: Bristol, UK, 2020; Volume 538. [CrossRef]
11. Pimpa, A.; Eiamkanitchat, N.; Phatsara, C.; Moonmanee, T. Decision support system for dairy cattle management using computational intelligence technique. In Proceedings of the 2019 7th International Conference on Computer and Communications Management, Bangkok, Thailand, 27–29 July 2019; pp. 181–185. [CrossRef]

12. Manabe, K.; Asami, Y.; Yamada, T.; Sugimori, H. Improvement in the Convolutional Neural Network for Computed Tomography Images. *Appl. Sci.* **2021**, *11*, 1505. [CrossRef]
13. Kabathova, J.; Drlik, M. Towards Predicting Student's Dropout in University Courses Using Different Machine Learning Techniques. *Appl. Sci.* **2021**, *11*, 3130. [CrossRef]
14. Choubin, B.; Abdolshahnejad, M.; Moradi, E.; Querol, X.; Mosavi, A.; Shamshirband, S.; Ghamisi, P. Spatial hazard assessment of the PM10 using machine learning models in Barcelona, Spain. *Sci. Total Environ.* **2020**, *701*, 134474. [CrossRef]
15. Šimić, I.; Lovrić, M.; Godec, R.; Kröll, M.; Bešlić, I. Applying machine learning methods to better understand, model and estimate mass concentrations of traffic-related pollutants at a typical street canyon. *Environ. Pollut.* **2020**, *263*, 114587. [CrossRef]
16. Grange, S.K.; Carslaw, D.C.; Lewis, A.C.; Boleti, E.; Hueglin, C. Random forest meteorological normalisation models for Swiss PM10 trend analysis. *Atmos. Chem. Phys.* **2018**, *18*, 6223–6239. [CrossRef]
17. Özdemir, U.; Taner, S. Impacts of Meteorological Factors on PM10: Artificial Neural Networks (ANN) and Multiple Linear Regression (MLR) Approaches. *Environ. Forensics* **2014**, *15*, 329–336. [CrossRef]
18. Cai, M.; Yin, Y.; Xie, M. Prediction of hourly air pollutant concentrations near urban arterials using artificial neural network approach. *Transp. Res. Part D Transp. Environ.* **2009**, *14*, 32–41. [CrossRef]
19. Biancofiore, F.; Busilacchio, M.; Verdecchia, M.; Aruffo, E.; Bianco, S.; Di Tommaso, S.; Colangeli, C.; Rosatelli, G.; Di Carlo, P. Recursive neural network model for analysis and forecast of PM10 and PM2.5. *Atmos. Pollut. Res.* **2017**, *8*, 652–659. [CrossRef]
20. Ceylan, Z.; Bulkan, S. Forecasting PM10 levels using ann and mlr: A case study for Sakarya city. *Glob. Nest J.* **2018**, *20*, 281–290. [CrossRef]
21. Goyal, P.; Chan, A.T.; Jaiswal, N. Statistical models for the prediction of respirable suspended particulate matter in urban cities. *Atmos. Environ.* **2006**, *40*, 2068–2077. [CrossRef]
22. Hamid, H.A.; Yahaya, A.S.; Ramli, N.A.; Ul-Saufie, A.Z.; Yasin, M.N. Short term prediction of PM10 concentrations using seasonal time series analysis. In *MATEC Web of Conferences*; EDP Sciences: Les Ulis, France, 2016; Volume 47, p. 05001. [CrossRef]
23. Fu, M.; Wang, W.; Le, Z.; Khorram, M.S. Prediction of particular matter concentrations by developed feed-forward neural network with rolling mechanism and gray model. *Neural Comput. Appl.* **2015**, *26*, 1789–1797. [CrossRef]
24. Perez, P. Combined model for PM10 forecasting in a large city. *Atmos. Environ.* **2012**, *60*, 271–276. [CrossRef]
25. Park, J.-H.; Yoo, S.-J.; Kim, K.-J.; Gu, Y.-H.; Lee, K.-H.; Son, U.-H. PM10 density forecast model using long short term memory. In Proceedings of the International Conference on Ubiquitous and Future Networks (ICUFN), Milan, Italy, 4–7 July 2017; pp. 576–581. [CrossRef]
26. Polat, K.; Durduran, S.S. Usage of output-dependent data scaling in modeling and prediction of air pollution daily concentration values (PM 10) in the city of Konya. *Neural Comput. Appl.* **2012**, *21*, 2153–2162. [CrossRef]
27. Yildirim, Y.; Bayramoglu, M. Adaptive Neuro-Fuzzy based modelling for prediction of air pollution daily levels in city of Zonguldak. *Chemosphere* **2006**, *63*, 1575–1582. [CrossRef] [PubMed]
28. Napook, P.; Eiamkanitchat, N. The adaptive dynamic clustering Neuro-Fuzzy system for classification. *Lect. Notes Electr. Eng.* **2015**, *339*, 721–728. [CrossRef]
29. Eiamkanitchat, N.; Theera-Umpon, N.; Auephanwiriyakul, S. A novel Neuro-Fuzzy method for linguistic feature selection and rule-based classification. In Proceedings of the International Conference on Computer and Automation Engineering (ICCAE), Singapore, 26–28 February 2010; pp. 247–252. [CrossRef]
30. Saetern, K.; Eiamkanitchat, N. An ensemble K-nearest neighbor with neuro-fuzzy method for classification. In Proceedings of the International Conference on Computing and Information Technology, (IC2IT), Phuket, Thailand, 8–9 May 2014. [CrossRef]
31. Eiamkanitchat, N.; Theera-Umpon, N.; Auephanwiriyakul, S. On Feature Selection and Rule Extraction for High Dimensional Data: A Case of Diffuse Large B-Cell Lymphomas Microarrays Classification. *Math. Probl. Eng.* **2015**, *2015*, 275831. [CrossRef]
32. Ross, T.J. Membership Functions, Fuzzification and Defuzzification. In *Fuzzy Systems in Medicine, Studies in Fuzziness and Soft Computing*; Physica-Verlag: Heidelberg, Germany, 2000; Volume 41, pp. 48–77. [CrossRef]
33. Thailand's Air Quality Information. Available online: air4thai.pcd.go.th/webV2/aqi_info.php (accessed on 7 November 2020).
34. Chaisornying, K.; Eiamkanitchat, N. Increasing Predictive Accuracy of Neuro-Fuzzy Using Quartiles to Initialize the Membership Function. In Proceedings of the International Conference on Computer and Communications Management (ICCCM), Singapore, 17–19 July 2020; pp. 130–133. [CrossRef]
35. Cheng, C.-H.; Chang, J.-R.; Yeh, C.-A. Entropy-based and trapezoid fuzzification-based fuzzy time series approaches for forecasting IT project cost. *Technol. Forecast. Soc. Chang.* **2006**, *73*, 524–542. [CrossRef]
36. Magaña-Villegas, E.; Carrera-Velueta, J.M.; Ramos-Herrera, S.; Hernández-Barajas, J.R.; González-Figueredo, C.; Laines-Canepa, J.R.; Valdés-Manzanilla, A.; Bautista-Margulis, R.G. Clustering approach applied on an artificial neural network model to predict PM10 in mega cities of Mexico. *Int. J. Sustain. Dev. Plan.* **2016**, *11*, 566–577. [CrossRef]
37. Srijiranon, K.; Eiamkanitchat, N. Collective Neural Networks System for PM10 Classification in the North of Thailand. In Proceedings of the 2018 22nd International Computer Science and Engineering Conference (ICSEC), Chiang Mai, Thailand, 21–24 November 2018; pp. 1–4. [CrossRef]
38. De Gennaro, G.; Trizio, L.; Di Gilio, A.; Pey, J.; Pérez, N.; Cusack, M.; Alastuey, A.; Querol, X. Neural network model for the prediction of PM10 daily concentrations in two sites in the Western Mediterranean. *Sci. Total Environ.* **2013**, *463–464*, 875–883. [CrossRef]

39. Sfetsos, A.; Vlachogiannis, D. An investigation of the effectiveness of advanced modeling tools on the forecasting of daily PM10 values in the Greater Athens area. In *Information Technologies in Environmental Engineering (ICSC)*; Springer: Berlin/Heidelberg, Germany, 2009; pp. 305–316. [CrossRef]
40. Vlachogianni, A.; Kassomenos, P.; Karppinen, A.; Karakitsios, S.; Kukkonen, J. Evaluation of a multiple regression model for the forecasting of the concentrations of NOx and PM10 in Athens and Helsinki. *Sci. Total Environ.* **2011**, *409*, 1559–1571. [CrossRef]
41. He, H.; Garcia, E.A. Learning from Imbalanced Data. *IEEE Trans. Knowl. Data Eng.* **2009**, *21*, 1263–1284. [CrossRef]
42. Chicco, D.; Jurman, G. The advantages of the Matthews correlation coefficient (MCC) over F1 score and accuracy in binary classification evaluation. *BMC Genom.* **2020**, *21*, 6. [CrossRef]

Article

Data Mining of Students' Consumption Behaviour Pattern Based on Self-Attention Graph Neural Network

Fangyao Xu † and Shaojie Qu *

Beijing Institute of Technology, Beijing 100081, China; 1120180084@bit.edu.cn
* Correspondence: qushaojie@bit.edu.cn
† Current address: No. 8, Liangxiang East Road, Fangshan District, Beijing 102488, China.

Abstract: Performance prediction is of significant importance. Previous mining of behaviour data was limited to machine learning models. Corresponding research has not made good use of the information of spatial location changes over time, in addition to discriminative students' behavioural patterns and tendentious behaviour. Thus, we establish students' behaviour networks, combine temporal and spatial information to mine behavioural patterns of academic performance discrimination, and predict student's performance. Firstly, we put forward some principles to build graphs with a topological structure based on consumption data; secondly, we propose an improved self-attention mechanism model; thirdly, we perform classification tasks related to academic performance, and determine discriminative learning and life behaviour sequence patterns. Results showed that the accuracy of the two-category classification reached 84.86% and that of the three-category classification reached 79.43%. In addition, students with good academic performance were observed to study in the classroom or library after dinner and lunch. Apart from returning to the dormitory in the evening, they tended to stay focused in the library and other learning venues during the day. Lastly, different nodes have different contributions to the prediction, thereby providing an approach for feature selection. Our research findings provide a method to grasp students' campus traces.

Keywords: self-attention mechanism; graph neural network; data mining; behaviour sequence pattern; behaviour network

Citation: Xu, F.; Qu, S. Data Mining of Students' Consumption Behaviour Pattern Based on Self-Attention Graph Neural Network. *Appl. Sci.* **2021**, *11*, 10784. https://doi.org/10.3390/app112210784

Academic Editor: Federico Divina

Received: 14 October 2021
Accepted: 12 November 2021
Published: 15 November 2021

Publisher's Note: MDPI stays neutral with regard to jurisdictional claims in published maps and institutional affiliations.

Copyright: © 2021 by the authors. Licensee MDPI, Basel, Switzerland. This article is an open access article distributed under the terms and conditions of the Creative Commons Attribution (CC BY) license (https://creativecommons.org/licenses/by/4.0/).

1. Introduction

Methods to improve education quality by mining off-line education and on-line learning platform data [1] has led to the development of educational data mining (EDM) [2]. Among the several problems in the field of EDM, predicting students' scholastic performance is a key issue [3,4], and various statistical methods [5,6] and tools [7] have been developed to perform this task. However, these methods could not reflect the learning conditions of students over a specific period, and do not facilitate the discovery of new knowledge patterns from the data set for the development of new and accurate models. Advancements in machine learning has led to the emergence of powerful data visualization methods and a variety of models, such as clustering, classification, and prediction, including algorithms that can dynamically process data streams [8], and other algorithms such as the support vector machine have been used to detect students who may fail in courses as an early warning [9]. Research on machine learning using behavioural data for performance prediction has attracted a significant amount of attention, such as grade prediction using online behaviour [10–12], gaming behaviour [13], consumption behaviour [14] and travel behaviour [15]. To conduct on-line behaviour mining more deeply, artificial neural networks have been used for log data mining and desirable results have been achieved [16]. However, with regard to knowledge tracking, such as the prediction of student's test questions, artificial neural networks have not yielded satisfactory result, and recurrent neural networks (RNNs) have had to be introduced [17]. Further, when dealing with long sequences, RNNs found that gradient explosion and gradient disappearance were prone to happen [18]. Therefore, to overcome such shortcomings, long short term memory networks (LSTM) were introduced [19]. LSTM can well solve the prediction problem with temporal data, while it has defects in processing spatial data. Further, the same behaviour will lead to ambiguity in the behavioural purpose if the difference in spatial location is ignored. For example, the behavioural meanings of fetching boiled water in the teaching building, in the dormitory, and in the bathroom are obviously different. We can, respectively, speculate that the purposes of corresponding behaviours are to study, to play games in the dormitory, and to take a bath. The lack of spatial location information will affect the prediction accuracy of models and the analysis of discriminative behavioural patterns. If the information in the spatial position was considered to mine behavioural characteristics contained in consumption data, the graph topological structure formed by behaviour cannot be well mined, and the above methods lost the ability to have good performance for such data with certain special structures. Naturally mining the data with graph structure for performance prediction requires the introduction of new tools. At present, graph neural networks (GNNs) have undergone rapid developments. Further, such networks have been extensively used [20–22] and are suitable for dealing with graph structure data. We observed that the behaviour of students can be demonstrated in the form of a graph and particular behavioural pattern and tendency can be observed, which inspired us to use graph-related tools to mine hidden information in behavioural patterns using students' spatial location changes over time.

In this study, we aimed to utilize the information regarding spatial position changes over time reflected by consumption data collected from a school to extract behavioural pattern features and construct graph structures for mining discriminative behavioural characteristics and behavioural trends related to academic performance, as well as for performance prediction. Firstly, we proposed two guidelines for constructing the graph structures by extracting features from the consumption behaviour data. Secondly, we proposed an improved self-attention mechanism model based on previous graph self-attention mechanism; thirdly, we made use of graphs composed of behaviour characteristics to perform classification and determined discriminative learning and life behaviour sequence patterns.

The remainder of this article is as follows: In Section 2, we present some additional current research on consumption behaviour and discuss the possibility of using a GNN for graph mining. In Section 3, we explain the data we used in our study as well as the processing methods; we also describe the method of graph construction and the improved model. In Section 4, we report the results of the experiment and analyse the results obtained. In Section 5, we propose some open issues, summarize the paper and also present some shortcomings of our research.

2. Related Work

Continuously adapting to rapid development and educational innovation is highly crucial and has led to the emergence and application of EDM in various fields. Scholars investigated the three aspects of student performance, teaching equality, and policy making, and found student performance had the greatest significance [23,24]. Related research had also focused on performance prediction and the discussion of methods [25] and models [26], including data [27], and models and methods were considerably improved for different purposes.

From the perspective of students, performance prediction [28,29], early warning of failure in subjects [30], sentiment analysis [31] and course recommendation [32,33] are key issues; besides, the trajectory of school behaviour is rich in information of learning habits of students to mine, and analyses of behavioural patterns based on spatial location change have been widely carried out for performance prediction. The trajectory of students' behaviour at school helps us understand the various characteristics of learning status and lifestyle [34]. Dalvi [35] focused on students' green and low-carbon behaviour, and Islam [36] studied electronic product consumption; in both these studies, only the lifestyles of students were investigated, and the impact of such consumption bahaviour on academic performance was not studied. Mei [37] used campus behaviour data to predict academic performance. However, discriminative learning and life patterns that can help distinguish students with different learning levels was unclear. Further, time and location information has not yet been considered in most studies. Li [14] focused on behaviours that could reflect the regularity of students' lives; however, behavioural patterns are not discriminative enough by nature. In study [38], based on the behaviour records of undergraduates' smart cards, the authors studied the impact of students' diligence and the regularity of their daily life on grades. Due to a lack of spatial information, the authors had to regard different behaviours as the same behaviour and the prediction results were affected. The main reason for this is the lack of consideration about spatial information. Thus, consideration of behavioural patterns with spatial location information is necessary.

Advanced statistical methods were extensively applied in EDM during its early stage, such as the *t*-test [39], for the prediction of academic performance [40,41]. Statistical methods were suitable for small sample data; otherwise, it was necessary to put forward intelligent algorithms. Machine learning algorithms were utilized for predicting academic performance [42] to warn students who might fail in certain courses [43] and predict the graduation rates [44]. Regarding machine learning algorithms that did not perform well on time series data, the improved recurrent neural network (RNN) based on artificial neural networks showed a good mining effect [45], and appeared to be ineffective for dealing with non-European data, such as those with a graph structure. There was also a lack of in-depth mining on student behaviour tendency reflected by the spatial location information. Due to the particularity of graph structure, it was also difficult to use general models to process and analyse. We therefore studied behavioural characteristics from the perspectives of spatial topological structure and time dimension, and we introduced a new and powerful tool for dealing with non-European structure data.

As a powerful tool for processing non-European structure data, graph neural networks have undergone rapid development and been wide applied, such as the knowledge graph [46], natural language processing [47], graph-based text representation [48] and graph embedding techniques [49]. In particular, some scholars have proposed the graph attention mechanism to improve the performance of node classification [50]. Some scholars have recommended graphs for recommendation systems [51], such as the music recommendation system in mobile networks [52,53] because of graphs' powerful information representation abilities and wide applications. Specifically, Zhang [54] used bipartite graph to perform context-sensitive web service discovery. Notably, community detection is also a key task [55,56]. Consumption behaviour at school has structure and characteristics similar to those of social networks and graph. A topology structure must therefore be introduced to distinguish behavioural patterns and find students' behaviour tendency. Inspired by the node classification method, we improved the present self-attention GNN to mine consumption behaviour data from both time and spatial aspects.

3. Methods

3.1. Data Description

As the behaviour of students is often the same regardless of the semester, we only collected behaviour data over a single month. The activities of first-year students are relatively messy due to their curiosity and are difficult to analyse, while the third-year and fourth-year students need to engage in some social work outside the school, which leads to very short time at school and consumption behaviours are lacking and inconvenient to analyse. Therefore, we considered the consumption data of all second-year students at Beijing Institute of Technology, which is characterized by science and engineering majors and has a male to female ratio of 2.2 to 1. Students use campus cards for on-campus consumption, and the corresponding data are transmitted and stored in the school's campus card consumption system later, which is relatively convenient to access. Thus, from December 2020 to January 2021, we collected 752,725 pieces of consumption data generated during May 2020 and 3640 pieces of final exam scores of the course of data structure from two data systems, including campus card consumption system and educational administration system. The collected consumption data are detailed in Table 1, and a more detailed explanation regarding the "action" is presented in Table 2.

Table 1. Collected raw data.

ID Number	Consumption Money (RMB Cent)	Time	Action
36984	200	2018/5/2 9:18	Breakfast
36984	400	2018/5/2 11:47	Lunch
36984	1500	2018/5/2 16:52	Dinner
17347	300	2018/5/31 11:24	Lunch
17347	250	2018/5/31 11:25	Lunch
17347	38	2018/5/31 11:26	Lunch
17347	150	2018/5/31 11:27	Lunch
10075	180	2018/5/20 11:16	Lunch
10075	200	2018/5/20 11:39	Lunch
10075	1500	2018/5/20 17:50	Dinner
10075	300	2018/5/21 07:56	Breakfast
10075	50	2018/5/21 07:56	Breakfast

Table 2. Specific explanation of different consumption behaviours.

Behaviour	Action Explanation
Dinner	Consumption after 4:00 p.m. in the cafeteria.
Lunch	Consumption between 10:00 a.m. and 4:00 p.m. in the cafeteria
Breakfast	Consumption before 10:00 a.m. in the cafeteria.
Supermarket	Consumption in the supermarket
Library	Consumption in the library
Dormitory bathroom	Consumption in the dormitory bathroom
Dormitory boiled water	Consumption on the dormitory water
Gym	Consumption in the school gym
School bus	Consumption caused by taking the school bus between campuses
Management office	Consumption in the school management office
Classroom boiled water	Consumption of water available in the classroom

3.2. Data Preprocessing

Based on the above data, we performed certain preprocessing measurements, using the following pseudo-code of Algorithm 1.

Algorithm 1 Data pre-processing

```
Input: Raw data file
Output: Processed data files for different students
Group data by ID
while Student ID equals some ID do
    if Two adjacent rows of data are exactly the same then
        Delete a row
    end if
    if The consumption behaviour is in the gym, management office or school bus then
        Delete the row
    end if
    if Two adjacent behaviours are the same then
        if Time difference is less than 5 fimnutes then
            Delete the second row
        end if
    end if
end while
if More than 15 consumption data for some student then
    Generate a new data file for the student
else
    Do not consider the student's behaviour data
end if
```

In collected data, "management office" refers to students' short-term campus card recharging behaviour; "school bus" refers to the behaviour of commuting between campuses, and the uncertainty of destination campus makes it impossible to analyse behavioural purpose; as for "gym", the overall time and purpose of staying in the gym based on this record cannot be inferred. Hence, we deleted related behaviours in the preprocessing algorithm. After the data preprocessing, we got the data of 3616 students and next considered extracting features from the following three aspects:

1. Indicators of regularity
2. Amount of consumption
3. Behaviour sequence pattern

After obtaining the features, we performed the chi-square test, f-test and other feature selection methods, and finally determined thirty relevant features related to the grades. Table 3 presents some of the information related to the features that were extracted.

We noted that the features related to consumption money were intermediate, which meant that excessive consumption and low consumption were both abnormal phenomena. Thus, this type of feature was transformed to the maximum feature using Formula (1):

$$x_i = 1 - \frac{|x_i - x_{\text{best}}|}{\max\{|x_i - x_{\text{best}}|\}} \tag{1}$$

The same method was applied to the other similar intermediate indicators, as presented in Table 3.

Table 3. Extracted features based on students' data and reality.

Type	Feature Name	Feature Explanation	Intermediate	Node ID
Indicators of regularity	Study-actual	Actual number of visits to library or classroom	No	0
	Paid-days	Number of days with consumption records	No	2
	Getting-up-num	Number of wake-ups	No	3
Amount of consumption	total-meals	Total number of meals	Yes	26
	Total-supermarket-num	Total number of times to go shopping	Yes	7
	Total-month-money	Total monthly cost	Yes	8
Behaviour sequence pattern	Dinner-dormitory-num	Number of times in the dormitory after dinner	No	13
	Lunch-study-num	Number of times to study after lunch	No	15
	Dinner-study-num	Number of times to study after dinner	No	16
	Dinner-study-dorm-bathroom-num	Number of times to study after dinner and go back to the dorm	No	21

3.3. Graph Construction

After the extraction of the behavioural sequence features, the data lost the natural graph structure. As such, construction of suitable graphs from these features became a challenge. To reflect the continuity and trend of students' behaviours, we constructed graphs based on the characteristics of students' behaviours at school and daily experience and regarded the features as nodes.

3.3.1. p-Clique

For a graph (V, E), V refers to a set of elements called vertices and E is a multiset of unordered pairs (u, v) whose elements are called edges. Two vertices are said to be adjacent if there exists an edge between them. A clique in a graph refers to a set of pairwise adjacent vertices, and if the number of vertices involved is p, it is called a p-clique. The behavioural patterns of students in their daily school life are often consistent. While some patterns may not be related to one another obviously, the extracted behaviour sequence features will be interrelated. This interrelationship between the features is an important indicator that can be used to distinguish between students with different learning levels. Based on this, we arranged part of the extracted behaviour sequence features into a p-clique in the constructed graph.

3.3.2. Other Criterion

Two nodes are said to be connected and interrelated if there exists an edge between them. Hence, for two nodes to have a relationship, we considered that the following two criteria must be fulfilled:

1. Necessary connection: This condition means that the two nodes are interrelated, for example, the edge between the total-month-money and total-lunch-money. The cost of lunch must be a part of the total monthly cost, and there exists a connection between the two features.
2. Unnecessary connection: This condition means that if the values of the two potentially related features are non-zero at the same time, there exists an edge between the two features. For example, if getting-up and breakfast-study are both non-zero at the same time, then there is a connection between the two features, which means that the student may tend to get up early for breakfast.

Based on the above-mentioned criterion, we constructed graphs for each student, creating a total of 3616 graphs for graph-level classification, using 700 graphs as the test set. Below, we present two graphs that were constructed according to the above-mentioned method (Figure 1).

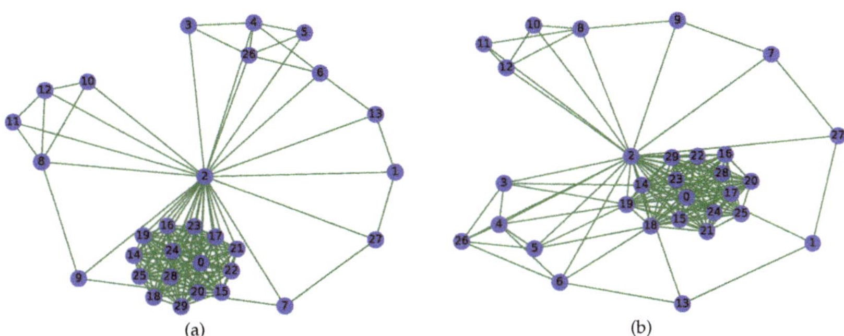

Figure 1. Different graph structures built for different students. The differences resulted from the students having different lifestyles. (**a**) Graph constructed for student with exam score 84.88 whose ID number equals 17347; (**b**) graph constructed for student with exam score 91.28 whose ID number equals 61465.

3.4. Model Description

We proposed an improved self-attention GNN based on previous research. The corresponding Algorithm 2 pseudo-code of improved self-attention model is as follows.

Algorithm 2 Graph classification based on improved self-attention GNN

Input: Adjacency matrix of constructed matrix W_{adj}, node feature vector V, graph indicators P and degree matrix D

Output: The prediction labels on the test set

1: Initialize: initialize weight matrix W and bias ε

2: Normalize adjacency matrix by $L = D^{-\frac{1}{2}}(W_{adj} + I)D^{-\frac{1}{2}}$, in which D refers to the degree matrix

3: Build a graph convolutional layer to obtain the attention scores by calculating $Z = GCN(L, V, W, \varepsilon)$

4: Perform convolution twice, use activation function ReLu to process the scores, and concatenate the scores S

5: Perform self-attention pooling by function pooling (W_{adj}, S, P) and update the graph structure and adjacency matrix W_{adj} according to the mask

6: Perform maximum pooling and average pooling, and concatenate the results

7: Predict the labels after the three fully connected layers

In the standard self-attention mechanism, given a group of nodes $(x_1, x_2, \cdots, x_k)$ and weight matrices W_Q, W_K, W_V that represent different linear transformations of features, the attention coefficients are computed to reflect the pair-wise importance of the nodes, as shown in Equation (2):

$$e_{ij} = \left(W_Q^T x_i\right)^T \left(W_K^T x_j\right), \forall 1 \le i, j \le k \tag{2}$$

Then e_{ij} is normalized by all possible values of j using the Softmax function as shown in Equation (3):

$$\alpha_{ij} = \frac{\exp(e_{ij})}{\sum_{1 \le l \le k} \exp(e_{il})} \tag{3}$$

Finally, a weighted sum of transformed features is calculated as shown in Equation (4):

$$\vec{d_i} = \tanh\left(\sum_{1 \le j \le k} \alpha_{ij} W_V^T x_j\right) \tag{4}$$

Additionally, a new node embedding vector set can be obtained when using a multi-head graph attention layer. In graph attention neural networks, the node used for the attention mechanism generally only aggregates the information of the first-order neighbours to update the information. In the improved model, due to the characteristics of the extracted features, and to utilize the neighbouring information of node v_i better, we applied two graph convolution layers. In the convolution layers, we calculated the node embedding using a linear transformer W and used the activation function ReLu to calculate the raw attention score between pair-wise nodes using Equations (5) and (6):

$$Z = GCN(L, V, W, \varepsilon) \tag{5}$$

$$e_{ij} = \text{ReLU}\left(\vec{a}^T \left(z_i^{(l)} \| z_j^{(l)}\right)\right) \tag{6}$$

where $\vec{a}$ is a weight vector for learning and $\|$ refers to concatenation. Next, we applied the SoftMax function to normalize e_{ij}, and a weighted sum based on attention on the features of all the neighbour nodes, as shown in Equation (7):

$$h_i^{(l+1)} = \text{ReLu}\left(\sum_{j\in\mathcal{N}(i)} \alpha_{ij}^{(l)} z_j^{(l)}\right) \tag{7}$$

$\mathcal{N}(i)$ refers to all the neighbours of node i. We also used the pooling method to handle redundant information and reduce the amount of calculations. The extracted features denote the occurrence frequency of consumption behaviour sequence patterns and the ReLu function shows the best performance on our dataset.

4. Result and Analysis

We compared our improved self-attention GNN to some other graph neural network variants on two classification methods, including some machine learning models. We will now discuss the difference in the loss reduction of the variant GCNs and the improved self-attention GNN mainly. Due to the uneven distribution of the labels, the results are weighted as demonstrated below.

4.1. Experiment Result

Since the proportion of students with a score of less than 60 was very small, we set a higher passing line. We used the score of 70 to divide the students' scores into two categories defined as $\{1, 2\}$ and conducted a training task. The student group whose scores are all greater than or equal to 70 is labeled 1, and the rest of the group whose scores are less than 70 is labeled 2. To construct a large graph and speed up calculations, we first batched all the training graphs, and then trained the self-attention GNN with 300 epochs, as shown in Figure 2. Compared with the other GNN variants trained using the same number of epochs, the loss of our improved model varied sharply during the training process. Table 4 lists the performance of different models.

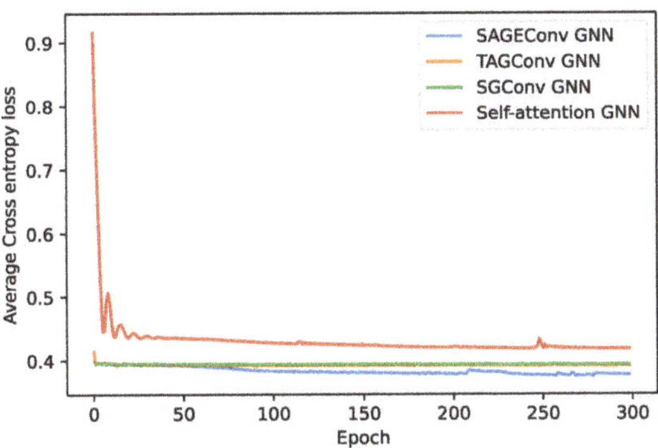

Figure 2. The processes aim at the two-category classification, in which "self-attention GNN" refers to our improved model.

Table 4. The two-category classification results on the test set.

Model	Accuracy	Precision	Recall	F1-Score
Improved self-attention graph neural network	84.86%	94.84%	84.86%	91.81%
Topology adaptive graph convolutional network (TAGConv GNN) [57]	68.57%	62.60%	68.57%	64.88%
GraphSAGE convolutional network (SAGEConv GNN) [58]	65.76%	65.29%	70.29%	67.04%
Simplified graph convolutional network (SGConv GNN) [59]	70.43%	66.19%	70.43%	67.71%
Logistic regression [60]	61.14%	81.39%	61.14%	66.64%
KNN [61]	84.00%	76.57%	84.00%	78.44%
Decision tree [62]	76.00%	77.00%	76.00%	77.00%

As can be seen from Figure 2, the cross-entropy loss reduction of the improved self-attention GNN was very fast, while it could quickly reach stability. While the descent process of the SAGEConv GNN exhibited fluctuations, the cross-entropy loss reduction was much smoother than our improved self-attention model. The remaining two graph neural network variants were both highly stable. This revealed the high sensitivity of our improved GNN model to variations in the data.

It was difficult to analyse the behavioural patterns of students who were outstanding in the two-category classification. Therefore, we further divided the students whose scores were greater than 70 into two categories and conducted a three-category classification task. Then, we used the scores of 70 and 85 to divide the students' scores into three categories defined as $\{1, 2, 3\}$ for multi-category classification. Specifically, students with a score of less than 70 are considered as failing, students with a score of 70 or above and not exceeding 85 are considered as good and students with a score of 85 or above are considered excellent, corresponding to label 1, label 2 and label 3, respectively. We then performed 300 epochs. We observed that the process of loss reduction gradually stabilized, as shown in Figure 3. The performances of different models are listed in Table 5.

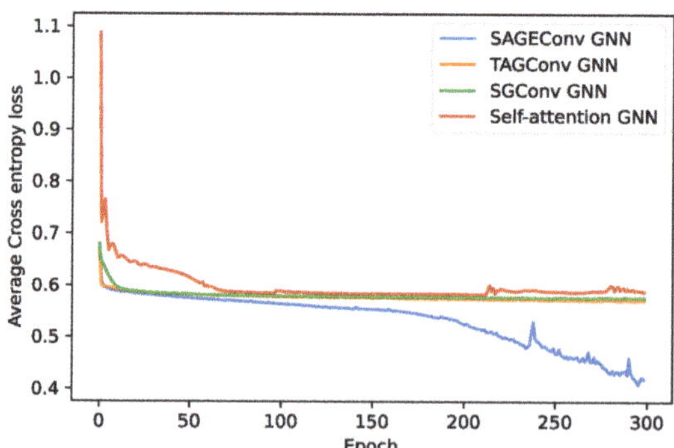

Figure 3. The processes aim at the three-category classification, in which "self-attention GNN" still means our improved model.

Table 5. The three-category classification results on the test set.

Model	Accuracy	Precision	Recall	F1-Score
Improved self-attention graph neural network	79.43%	97.20%	79.43%	87.28%
Topology adaptive graph convolutional network (TAGConv GNN) [57]	64.71%	62.44%	64.71%	62.99%
GraphSAGE convolutional network (SAGEConv GNN) [58]	66.06%	65.91%	65.86%	65.43%
Simplified graph convolutional network (SGConv GNN) [59]	67.00%	63.51%	67.00%	64.56%
Logistic regression [60]	48.29%	72.80%	48.29%	53.49%
KNN [61]	79.14%	68.93%	79.14%	68.00%
Decision tree [62]	68.05%	67.16%	68.09%	68.17%

The improved self-attention GNN showed a high degree of data sensitivity from Figure 3; after reaching stability, the cross-entropy loss of our model still showed some fluctuations. After training SAGEConv GNN for 300 epochs, the cross-entropy loss kept on declining and fluctuating from the declining process of the cross entropy loss. Therefore, our improved self-attention GNN was better than SAGEConv GNN in this regard, which implied that the two models may have high data requirements and data sensitivity.

Based on the idea of hypothesis testing, we constructed an indicator based on the discriminative behavioural patterns to reflect the differences in behavioural patterns using Formula (8):

$$\text{ratio} = \frac{\sum_{i=1}^{n} x_i / n}{\sum_{j=1}^{m} y_j / m} \tag{8}$$

In the two-category classification, x represents the number of occurrences of some behavioural patterns of outstanding students, while y represents the number of occurrences of corresponding behavioural pattern of lagging students; in the three-category classifications, three ratio indicators were used regarding three score categories, respectively. The numerator and denominator were divided by the corresponding number of people to make sure the number of people did not have an impact. Taking the two-category classification as an example, the related results are presented in Table 6.

4.2. Result Analysis

From the perspective of cross entropy loss decreasing process in Figures 2 and 3, compared with other models, the improved self-attention model had higher cross-entropy loss in the initial stage of training, while with certain fluctuations, it could decrease and converge rapidly in both classification tasks. The initial cross-entropy loss and training processes of the other three graph neural network model variants were similar in two-category classification, which may be related to the fact that these models treated the nodes indiscriminately during each iteration. However, the loss of SAGEConv GNN maintained the downward trend in three-category classification. Judging from the model's prediction accuracy and other performance, we could speculate that although the loss declined, the model might have been over-fitting, which led to poor performance on the test set. In addition, the training process of both our improved model and SAGEConv GNN demonstrated larger fluctuations, which revealed that they might be more sensitive to label distribution.

Table 6. Ratio of occurrences of different discriminative behavioural patterns in the two category classification tasks.

Behavioural Pattern Sequence	Average Times of Outstanding Students	Average Times of Lagging Students	Ratio	Node ID
Study-actual	5.3516	2.2573	2.3708	0
Paid-days	29.5463	27.5556	1.0722	2
Getting-up-num	10.7108	4.9357	2.1701	3
Dinner-dormitory-num	8.3119	6.0819	1.3667	13
Breakfast-study-num	1.2968	0.4269	3.0377	14
Lunch-study-num	1.0605	0.3275	3.2383	15
Dinner-study-num	0.7278	0.2982	2.4402	16
Classroom-lunch-or-dinner-num	1.8204	0.8655	2.1033	18
Breakfast-study-lunch-num	0.7675	0.2690	2.8530	19
Dinner-study-dorm-bathroom-num	0.2287	0.0526	4.4359	21
Total-meals	56.3516	42.7252	1.3189	26

From the perspective of prediction accuracy in Tables 4 and 5, the improved self-attention GNN performed better than any other model in both the classification tasks and achieved accuracy of 84.86% in two-category classification and 79.43% in three-category classification, respectively. Moreover, precision could reach 94.84% and 97.20%, and F1-score was able to achieve 91.81% and 87.28%, respectively. Indicators, such as precision, recall score and F1-score, revealed that improving the self-attention model could improve the prediction performance and converge to stability rapidly. Judging from several indicators, three variants performed similarly but worse than our improved model, in which the accuracy reached 68.57%, 65.76%, 70.43% and 64.71%, 66.06%, 67% in the two tasks, respectively. They focused on improving graph convolution methods. Compared with machine learning algorithms, the overall performance of the graph neural network variants was found to be superior, demonstrating the graph's strong and in-depth ability to mine mutual influences between local nodes. As comparison, the machine learning models equalized the input features, ignoring their potential mutual influence, which resulted in their relatively unsatisfactory performance in some extent. The KNN classification performance was also relatively good, in which accuracy and recall rate were the same, reaching 84% and 79.14% in the two tasks, respectively. However, the KNN could not predict the behavioural trend and identify discriminative behavioural patterns. Its performance was dependent on the training set and the way the distance between data points was defined, which limited its applicability. The decision tree can give judgments through internal nodes and determine classification results based on leaf nodes; however, the depth and the number of leaf nodes of the decision tree for good predictive effect had certain uncertainty and it could not predict students' behavioural tendency similar to the KNN. It was more suitable for ensemble learning to improve the accuracy, while in this research accuracy only achieved 76% and 68.05% in two tasks and bore a resemblance to the remaining three indicators. As for logistic regression, this method could be regarded as generalized linear regression, while a more complex non-linear relationship cannot be expressed, not to mention mining local mutual influence. The above shortcomings contributed to such a situation: Accuracy equaled to recall rate in both tasks by 61.14% and 48.29%, precisions were 81.39% and 72.8% and F1-scores were 66.64% and 53.49%. By paying more attention to the local structure information using the GNN, better local mutual information mining could be performed, except improving the prediction accuracy of our method. The improved self-attention GNN could further identify discriminative behavioural patterns and predict the behavioural trends of students from the node scores and the existence of edges, which provided an insight into the behavioural patterns of outstanding students.

Next we discussed the prediction of students' behavioural trends. Because of the use of the self-attention mechanism, different feature nodes contributed to the performance prediction differently. We utilize different colour shades to represent the scores of the different nodes as shown in Figure 4. The darker the colour, the higher the score, demonstrating the relationship between the different behavioural patterns and their contribution to predicting the academic performance of the student. The first figure (Figure 4a) represents one of the graph trainings resulting from the two-category classification task, followed by three-category classifications. In two different classification tasks, different nodes showed different contributions for grade prediction.

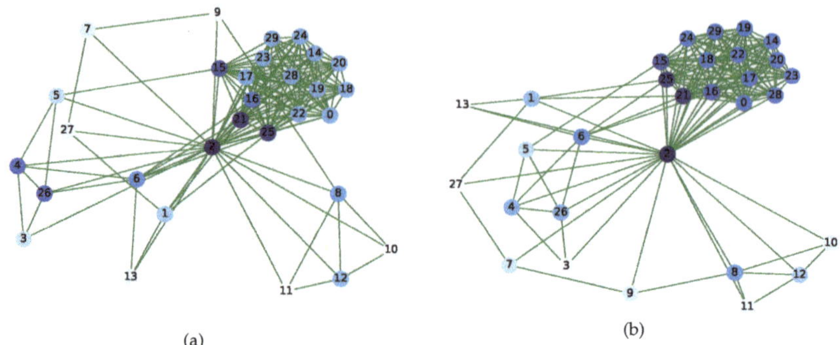

Figure 4. The score of the graph node after improved self-attention GNN training is indicated in each node by colour shade. (**a**) Two-category classification for students whose ID number is 61465 and exam score is 91.28; (**b**) Three-category classification for student whose ID number is 61465 and exam score is 91.28.

Among them, the paid-days (node 2) had the highest score, as can be seen from Figure 4. The results showed that daily life habits of the student were relatively consistent and regular. In addition, we noticed that the behavioural pattern of studying after dinner and then returning to the dormitory (node 25), and the number of times to study after lunch (node 15) influenced the prediction greatly. In addition, we noticed that whether the student at least ate lunch and dinner during the day impacted on their academic performance; this behavioural pattern was also a reflection of the regularity in their daily life. Additionally, taking the graph shown in Figure 4a as an example and considering the behavioural tendency, it was most likely that the student would study and then return to the dormitory after finishing dinner according to single node 21. Among the neighbours of node 4 (number of times eating both breakfast and lunch or both lunch and dinner), the influence of node 6 (number of dinners) was relatively large, which showed that if the student tended to enjoy dinner, the student was most likely to eat all three meals. It is therefore possible to create a similar analysis for other students as well.

Taking discriminative behavioural patterns into consideration, in Table 6, except for following behavioural patterns, such as the number of days with consumption records, returning to the dormitory after dinner, and the total number of meals over a month, the ratios of the remaining indicators exceeded 2. On average, the number of occurrences of these behavioural patterns of outstanding students exceeded twice that of lagging students. Further, we speculated that some corresponding behavioural patterns of some top students occurred more frequently. The visualization results were consistent with the above-mentioned results, thereby proving the correctness and rationality of our conclusions. Therefore, we believe that these characteristics could be used as a reference to distinguish students with regard to their learning levels.

5. Open Issues and Conclusions

5.1. Open Issues

In fact, judging from the results, GNN still needs to be improved in EDM. Below we list some open issues for further research:

1. How to use nodes and edges to reflect the relationship between features is still an open question.
2. The learning data is various, such as the data generated during MOOC online course learning, which inspires us to construct graphs to present more kinds of data for information mining.

3. As for node score based on self-attention mechanism, the features represented by different nodes can be ranked by importance. Thus, how to combine traditional feature selection methods and GNN to determine useful features is also worth exploring.
4. Through the comparison of scores, we can judge what the student's next behaviour pattern is most likely to be. However, based on the score values obtained from training, new quantification methods are needed.
5. How to build a behavioural network of all students, and integrate more information including curriculum arrangements, to conduct more diversified analysis, including skipping classes and social interactions, is encouraged.

5.2. Conclusions

In this study, we focused on mining on-campus consumption data to identify discriminative behavioural patterns based on spatial location change, analyse behaviour trends, and predict students' academic performance by constructing behaviour networks. Firstly, we preprocessed collected consumption data to extract features that could reflect students' living habits and their learning status. Secondly, we attempted to construct campus behaviour networks from reality and artificially, including introducing p-clique. Thirdly, combining with the pooling method, an improved self-attention GNN was utilized for training and prediction, and good prediction performance on the test set was achieved. For discriminative behavioural patterns, the habit of getting up early and behavioural patterns of continuous learning until meal time (lunch or dinner) in classroom and learning after three meals were discriminatory for distinguishing students with different learning levels, since the defined ratios of these behavioural pattern features are greater than 2 and we held certain beliefs that these characteristics are discriminatory in the sense of average. These new knowledge discoveries of behavioural patterns were consistent with the visualization results, conformed to the actual situation, and had certain reference meaning. For the behavioural trend analysis, judging from the behavioural pattern represented by single node 21, the student was more likely to continue studying after dinner than to return to the dormitory; from the perspective of node interaction and the existence of edges (e.g., node 4 and its neighbour nodes), three meals of students who often eat dinner were also relatively regular.

However, some limitations need noting regarding the present study. An arguable weakness is that all the graphs we consider are still not large samples. When we consider multi-category classification, the current method may fall into the shortcomings of few training samples for each category. In the meantime, few-shot learning may contribute to solving the above problem. Another weakness is that we discard some information, such as excluding behaviour related to gym and school busses. In addition, application limitation exists in our model and it is hard to apply in other universities.

As for future research directions, we were considering building a bigger behavioual network involving all students at school by their consumption record and some useful and possible video material to carry out analyses for different purposes, such as detecting absenteeism.

Author Contributions: Conceptualization, S.Q. and F.X.; methodology, F.X.; software, F.X.; validation, S.Q.; formal analysis, F.X. and S.Q.; investigation, S.Q.; resources, S.Q.; data curation, F.X.; writing—original draft preparation, F.X.; writing—review and editing, S.Q.; visualization, F.X.; supervision, S.Q.; project administration, S.Q.; funding acquisition, S.Q. All authors have read and agreed to the published version of the manuscript.

Funding: This research received no external funding.

Institutional Review Board Statement: Not applicable.

Informed Consent Statement: Not applicable.

Data Availability Statement: The data presented in this study are available on request from the corresponding author. The data are not publicly available due to privacy, including code.

Conflicts of Interest: The authors declare that there is no conflict of interest regarding the publication of this paper.

References

1. Jugo, I.; Kovai, B.; Slavuj, V. Increasing the adaptivity of an intelligent tutoring system with educational data mining: A system overview. *Int. J. Emerg. Technol. Learn.* **2016**, *11*, 67–70. [CrossRef]
2. Grigorova, K.; Malysheva, E.; Bobrovskiy, S. Application of Data Mining and Process Mining approaches for improving e-Learning Processes. In Proceedings of the 3rd International Conference on Information Technology and Nanotechnology, Samara, Russia, 24–27 April 2017; Volume 1903, pp. 115–121. [CrossRef]
3. Karthikeyan, V.G.; Thangaraj, P.; Karthik, S. Towards developing hybrid educational data mining model (HEDM) for efficient and accurate student performance evaluation. *Soft Comput.* **2020**, *24*, 18477–18487. [CrossRef]
4. Anoopkumar, M.; Md Zubair Rahman, A. A Review on Data Mining techniques and factors used in Educational Data Mining to predict student amelioration. In Proceedings of the 2016 International Conference on Data Mining and Advanced Computing (SAPIENCE), Ernakulam, India, 16–18 March 2016; pp. 122–133. [CrossRef]
5. Fernandes, E.; Carvalho, R.; Holanda, M.; Van Erven, G. Educational data mining: Discovery standards of academic performance by students in public high schools in the federal district of Brazil. In *World Conference on Information Systems and Technologies*; Springer: Cham, Switzerland, 2017; Volume 569, pp. 287–296. [CrossRef]
6. Nuankaew, W.; Nuankaew, P.; Teeraputon, D.; Phanniphong, K.; Bussaman, S. Perception and attitude toward self-regulated learning of Thailand's students in educational data mining perspective. *Int. J. Emerg. Technol. Learn.* **2019**, *14*, 34–49. [CrossRef]
7. Sabourin, J.; McQuiggan, S.; de Waal, A. SAS Tools for educational data mining. In Proceedings of the EDM 2016, Raleigh, NC, USA, 29 June–2 July 2016; pp. 632–633.
8. Xu, S.; Wang, J. Dynamic extreme learning machine for data stream classification. *Neurocomputing* **2017**, *238*, 433–449. [CrossRef]
9. Costa, E.B.; Fonseca, B.; Santana, M.A.; de Araujo, F.F.; Rego, J. Evaluating the effectiveness of educational data mining techniques for early prediction of students academic failure in introductory programming courses. *Comput. Hum. Behav.* **2017**, *73*, 247–256. [CrossRef]
10. Ducange, P.; Pecori, R.; Sarti, L.; Vecchio, M. Educational big data mining: How to enhance virtual learning environments. In *International Conference on EUropean Transnational Education*; Springer: Berlin/Heidelberg, Germany, 2017; Volume 527, pp. 681–690. [CrossRef]
11. Chen, J.; Zhao, J. An educational data mining model for supervision of network learning process. *Int. J. Emerg. Technol. Learn.* **2018**, *13*, 67–77. [CrossRef]
12. de J. Costa, J.; Bernardini, F.; Artigas, D.; Viterbo, J. Mining direct acyclic graphs to find frequent substructures—An experimental analysis on educational data. *Inf. Sci.* **2019**, *482*, 266–278. Available online: https://www.sciencedirect.com/science/article/pii/S0020025519300398 (accessed on 11 January 2019). [CrossRef]
13. Malkiewich, L.; Baker, R.S.; Shute, V.; Kai, S.; Paquette, L. Classifying behaviour to elucidate elegant problem solving in an educational game. In Proceedings of the Ninth International Conference on Educational Data Mining, Raleigh, NC, USA, 29 June–2 July 2016; pp. 448–453.
14. Li, Y.; Li, D. University students' behaviour characteristics analysis and prediction method based on combined data mining model. In Proceedings of the 2020 3rd International Conference on Computers in Management and Business, Tokyo, Japan, 31 January–2 February 2020; pp. 9–13. [CrossRef]
15. Zheng, L.; Xia, D.; Zhao, X.; Tan, L.; Li, H.; Chen, L.; Liu, W. Spatial–temporal travel pattern mining using massive taxi trajectory data. *Phys. A Stat. Mech. Its Appl.* **2018**, *501*, 24–41. [CrossRef]
16. Altaf, S.; Soomro, W.; Rawi, M.I.M. Student Performance Prediction using Multi-Layers Artificial Neural Networks: A case study on educational data mining. In Proceedings of the 2019 3rd International Conference on Information System and Data Mining, Houston, TX, USA, 6–8 April 2019; pp. 59–64. [CrossRef]
17. Nakagawa, H.; Iwasawa, Y.; Matsuo, Y. End-to-end deep knowledge tracing by learning binary question-embedding. In Proceedings of the 2018 IEEE International Conference on Data Mining Workshops (ICDMW), Singapore, 17–20 November 2018; pp. 334–342. [CrossRef]
18. Pascanu, R.; Mikolov, T.; Bengio, Y. On the difficulty of training recurrent neural networks. In Proceedings of the International Conference on Machine Learning, Atlanta, GA, USA, 17–19 June 2013; Number PART 3, pp. 2347–2355.
19. Tseng, C.W.; Chou, J.J.; Tsai, Y.C. Text mining analysis of teaching evaluation questionnaires for the selection of outstanding teaching faculty members. *IEEE Access* **2018**, *6*, 72870–72879. [CrossRef]
20. Morsy, S.; Karypis, G. A study on curriculum planning and its relationship with graduation GPA and time to degree. In Proceedings of the 9th International Conference on Learning Analytics & Knowledge, Tempe, AZ, USA, 4–8 March 2019; pp. 26–35. [CrossRef]
21. Hu, Q.; Polyzou, A.; Karypis, G.; Rangwala, H. Enriching course-Specific regression models with content features for grade prediction. In Proceedings of the 2017 IEEE International Conference on Data Science and Advanced Analytics (DSAA), Tokyo, Japan, 19–21 October 2017; Volume 2018, pp. 504–513. [CrossRef]
22. Yang, Y.; Liu, H.; Carbonell, J.; Ma, W. Concept graph learning from educational data. In Proceedings of the Eighth ACM International Conference on Web Search and Data Mining, Shanghai, China, 2–6 February 2015; pp. 159–168. [CrossRef]

23. Aldowah, H.; Al-Samarraie, H.; Fauzy, W.M. Educational data mining and learning analytics for 21st century higher education: A review and synthesis. *Telemat. Inform.* **2019**, *37*, 13–49. [CrossRef]
24. Jones, K.M.; Rubel, A.; LeClere, E. A matter of trust: Higher education institutions as information fiduciaries in an age of educational data mining and learning analytics. *J. Assoc. Inf. Sci. Technol.* **2020**, *71*, 1227–1241. [CrossRef]
25. Amrieh, E.A.; Hamtini, T.; Aljarah, I. Mining educational data to predict student's academic performance using ensemble methods. *Int. J. Database Theory Appl.* **2016**, *9*, 119–136. [CrossRef]
26. Bhagavan, K.S.; Thangakumar, J.; Subramanian, D.V. Predictive analysis of student academic performance and employability chances using HLVQ algorithm. *J. Ambient Intell. Humaniz. Comput.* **2021**, *12*, 3789–3797. [CrossRef]
27. Gao, H.; Qi, G.; Ji, Q. Schema induction from incomplete semantic data. *Intell. Data Anal.* **2018**, *22*, 1337–1353. [CrossRef]
28. Wang, X.; Yu, X.; Guo, L.; Liu, F.; Xu, L. Student performance prediction with short-term sequential campus behaviours. *Information* **2020**, *11*, 201. [CrossRef]
29. Wu, Z.; He, T.; Mao, C.; Huang, C. Exam paper generation based on performance prediction of student group. *Inf. Sci.* **2020**, *532*, 72–90. Available online: https://www.sciencedirect.com/science/article/pii/S0020025520303716 (accessed on 4 May 2020). [CrossRef]
30. Sun, Y.; Chai, R. An early-warning model for online learners based on user portrait. *Ing. Des Syst. D'Inf.* **2020**, *25*, 535–541. [CrossRef]
31. Onan, A. Sentiment analysis on massive open online course evaluations: A text mining and deep learning approach. *Comput. Appl. Eng. Educ.* **2021**, *29*, 572–589. [CrossRef]
32. Zhang, H.; Huang, T.; Lv, Z.; Liu, S.; Zhou, Z. MCRS: A course recommendation system for MOOCs. *Multimed. Tools Appl.* **2018**, *77*, 7051–7069. [CrossRef]
33. Kardan, A.A.; Ebrahimi, M. A novel approach to hybrid recommendation systems based on association rules mining for content recommendation in asynchronous discussion groups. *Inf. Sci.* **2013**, *219*, 93–110. Available online: https://www.sciencedirect.com/science/article/pii/S0020025512004756 (accessed on 24 July 2012). [CrossRef]
34. Xie, T.; Zheng, Q.; Zhang, W. Mining temporal characteristics of behaviours from interval events in e-learning. *Inf. Sci.* **2018**, *447*, 169–185. Available online: https://www.sciencedirect.com/science/article/pii/S0020025518301993 (accessed on 1 June 2018). [CrossRef]
35. Dalvi-Esfahani, M.; Alaedini, Z.; Nilashi, M.; Samad, S.; Asadi, S.; Mohammadi, M. Students green information technology behaviour: Beliefs and personality traits. *J. Clean. Prod.* **2020**, *257*, 120406. [CrossRef]
36. Islam, M.T.; Dias, P.; Huda, N. Young consumers e-waste awareness, consumption, disposal, and recycling behaviour: A case study of university students in Sydney, Australia. *J. Clean. Prod.* **2021**, *282*, 124490. [CrossRef]
37. Mei, G.; Hou, Y.; Zhang, T.; Xu, W. Behaviour Represents Achievement: Academic Performance Analytics of Engineering Students via Campus Data. In *2020 Chinese Automation Congress (CAC)*; IEEE: Piscataway, NJ, USA, 2020; pp. 4348–4353. [CrossRef]
38. Cao, Y.; Gao, J.; Lian, D.; Rong, Z.; Shi, J.; Wang, Q.; Wu, Y.; Yao, H.; Zhou, T. Orderliness predicts academic performance: Behavioural analysis on campus lifestyle. *J. R. Soc. Interface* **2018**, *15*, 20180210. [CrossRef]
39. Vijayalakshmi, M.; Salimath, S.; Shettar, A.S.; Bhadri, G. A study of team formation strategies and their impact on individual student learning using educational data mining (EDM). In Proceedings of the 2018 IEEE Tenth International Conference on Technology for Education (T4E), Chennai, India, 10–13 December 2018; pp. 182–185. [CrossRef]
40. Hao, J.; Liu, L.; von Davier, A.A.; Kyllonen, P.; Kitchen, C. *Collaborative Problem Solving Skills versus Collaboration Outcomes: Findings from Statistical Analysis and Data Mining*; International Educational Data Mining Society: Raleigh, NC, USA, 2016; pp. 382–387.
41. Gowri, G.; Thulasiram, R.; Baburao, M.A. Educational Data Mining Application for Estimating Students Performance in Weka Environment. In *IOP Conference Series: Materials Science and Engineering*; IOP Publishing: Bristol, UK, 2017; Volume 263. [CrossRef]
42. Jovanovic, M.; Vukicevic, M.; Milovanovic, M.; Minovic, M. Using data mining on student behaviour and cognitive style data for improving e-learning systems: A case study. *Int. J. Comput. Intell. Syst.* **2012**, *5*, 597–610. [CrossRef]
43. Viloria, A.; Garcia Guliany, J.; Niebles Nuz, W.; Hernandez Palma, H.; Niebles Nuz, L. Data Mining Applied in School Dropout Prediction. *J. Phys. Conf. Ser.* **2020**, *1432*. [CrossRef]
44. Injadat, M.; Moubayed, A.; Nassif, A.B.; Shami, A. Multi-split optimized bagging ensemble model selection for multi-class educational data mining. *Appl. Intell.* **2020**, *50*, 4506–4528. [CrossRef]
45. Matayoshi, J.; Cosyn, E.; Uzun, H. Are We There Yet? Evaluating the Effectiveness of a Recurrent Neural Network-Based Stopping Algorithm for an Adaptive Assessment. *Int. J. Artif. Intell. Educ.* **2021**, *31*, 304–336. [CrossRef]
46. Issa, S.; Adekunle, O.; Hamdi, F.; Cherfi, S.S.S.; Dumontier, M.; Zaveri, A. Knowledge Graph Completeness: A Systematic Literature Review. *IEEE Access* **2021**, *9*, 31322–31339. [CrossRef]
47. Vashishth, S.; Yadati, N.; Talukdar, P. Graph-based deep learning in natural language processing. In Proceedings of the 7th ACM IKDD CoDS and 25th COMAD, Hyderabad, India, 5–7 January 2020; pp. 371–372. [CrossRef]
48. Osman, A.H.; Barukub, O.M. Graph-Based Text Representation and Matching: A Review of the State of the Art and Future Challenges. *IEEE Access* **2020**, *8*, 87562–87583. [CrossRef]
49. Chen, Y.; Wu, Y.; Ma, S.; King, I. A Literature Review of Recent Graph Embedding Techniques for Biomedical Data. In *International Conference on Neural Information Processing 2020*; Springer: Cham, Switzerland, 2020; Volume 1333, pp. 21–29. [CrossRef]
50. Veličković, P.; Cucurull, G.; Casanova, A.; Romero, A.; Lio, P.; Bengio, Y. Graph attention networks. *arXiv* **2017**, arXiv:1710.10903.

51. Kherad, M.; Bidgoly, A.J. Recommendation system using a deep learning and graph analysis approach. *arXiv* **2020**, arXiv:2004.08100.
52. Wang, R.; Ma, X.; Jiang, C.; Ye, Y.; Zhang, Y. Heterogeneous information network-based music recommendation system in mobile networks. *Comput. Commun.* **2020**, *150*, 429–437. [CrossRef]
53. Durand, G.; Belacel, N.; LaPlante, F. Graph theory based model for learning path recommendation. *Inf. Sci.* **2013**, *251*, 10–21. Available online: https://www.sciencedirect.com/science/article/pii/S0020025513003149 (accessed on 30 April 2013). [CrossRef]
54. Zhang, R.; Zettsu, K.; Kidawara, Y.; Kiyoki, Y.; Zhou, A. Context-sensitive Web service discovery over the bipartite graph model. *Front. Comput. Sci.* **2013**, *7*, 875–893. [CrossRef]
55. Zhao, X.; Liang, J.; Wang, J. A community detection algorithm based on graph compression for large-scale social networks. *Inf. Sci.* **2021**, *551*, 358–372. [CrossRef]
56. Chen, J.; Li, R.; Zhao, S.; Zhang, Y.P. A New Clustering Cover Algorithm Based on Graph Representation for Community Detection. *Tien Tzu Hsueh Pao/Acta Electron. Sin.* **2020**, *48*, 1680–1687. [CrossRef]
57. Du, J.; Zhang, S.; Wu, G.; Moura, J.M.; Kar, S. Topology adaptive graph convolutional networks. *arXiv* **2017**, arXiv:1710.10370.
58. Hamilton, W.L.; Ying, R.; Leskovec, J. Inductive representation learning on large graphs. In Proceedings of the 31st International Conference on Neural Information Processing Systems, Long Beach, CA, USA, 4–9 December 2017; pp. 1025–1035.
59. Wu, F.; Souza, A.; Zhang, T.; Fifty, C.; Yu, T.; Weinberger, K. Simplifying graph convolutional networks. In Proceedings of the International Conference on Machine Learning, Long Beach, CA, USA, 10–15 June 2019; pp. 6861–6871.
60. Cuji Chacha, B.R.; Gavilanes Lopez, W.L.; Vicente Guerrero, V.X.; Villacis Villacis, W.G. Student Dropout Model Based on Logistic Regression. In *International Conference on Applied Technologies 2020*; Springer: Cham, Switzerland, 2020; Volume 1194, pp. 321–333. [CrossRef]
61. Dervisevic, O.; Zunic, E.; Donko, D.; Buza, E. Application of KNN and Decision Tree Classification Algorithms in the Prediction of Education Success from the Edu720 Platform. In Proceedings of the 2019 4th International Conference on Smart and Sustainable Technologies (SpliTech), Split, Croatia, 18–21 June 2019. [CrossRef]
62. Mkwazu, H.R.; Yan, C. Grade Prediction Method for University Course Selection Based on Decision Tree. In Proceedings of the 2020 International Conference on Aviation Safety and Information Technology, Weihai, China, 14–16 October 2020; pp. 593–599. [CrossRef]

Article

Deep Learning-Based Water Crystal Classification

Hien Doan Thi [1,†], Frederic Andres [2,*,†], Long Tran Quoc [1,†], Hiro Emoto [3,†], Michiko Hayashi [3,†], Ken Katsumata [3,†] and Takayuki Oshide [3,†]

1 Department of Computer Science, Vietnam National University, Hanoi 11300, Vietnam; hiendt104@gmail.com (H.D.T.); tqlong@vnu.edu.vn (L.T.Q.)
2 National Institute of Informatics, Tokyo 101-8430, Japan
3 I.H.M General Research Institute, Tokyo 103-0004, Japan; hiromasa@hado.com (H.E.); hayashi@hado.com (M.H.); katsumata@hado.com (K.K.); oside@hado.com (T.O.)
* Correspondence: andres@nii.ac.jp; Tel.: +81-3-4212-2542
† These authors contributed equally to this work.

Abstract: Much of the earth's surface is covered by water. As was pointed out in the 2020 edition of the World Water Development Report, climate change challenges the sustainability of global water resources, so it is important to monitor the quality of water to preserve sustainable water resources. Quality of water can be related to the structure of water crystal, the solid-state of water, so methods to understand water crystals can help to improve water quality. As a first step, a water crystal exploratory analysis has been initiated with the cooperation with the Emoto Peace Project (EPP). The 5K EPP dataset has been created as the first world-wide small dataset of water crystals. Our research focused on reducing the inherent limitations when fitting machine learning models to the 5K EPP dataset. One major result is the classification of water crystals and how to split our small dataset into several related groups. Using the 5K EPP dataset of human observations and past research on snow crystal classification, we created a simple set of visual labels to identify water crystal shapes, in 13 categories. A deep learning-based method has been used to automatically do the classification task with a subset of the label dataset. The classification achieved high accuracy when using a fine-tuning technique.

Keywords: water crystal; deep learning; fine-tuning; supervised; classification

Citation: Thi, H.D.; Andres, F.; Quoc, L.T.; Emoto, H.; Hayashi, M.; Katsumata, K.; Oshite. T. Deep Learning-Based Water Crystal Classification. *Appl. Sci.* **2022**, *12*, 825. https://doi.org/10.3390/app12020825

Academic Editor: Chuan-Ming Liu

Received: 31 October 2021
Accepted: 10 December 2021
Published: 14 January 2022

Publisher's Note: MDPI stays neutral with regard to jurisdictional claims in published maps and institutional affiliations.

Copyright: © 2022 by the authors. Licensee MDPI, Basel, Switzerland. This article is an open access article distributed under the terms and conditions of the Creative Commons Attribution (CC BY) license (https://creativecommons.org/licenses/by/4.0/).

1. Introduction

Along with the development of society, research on the human impact on nature is of greater and greater concern. Water quality [1] has become one of the main challenges that societies will face during the 21st century, as the United Nations brought water quality issues to the forefront of international actions under the Sustainable Development Goal 6. It is important to monitor how human actions will affect water quality and pollution issues. Water has always played an important role in the climatic ecosystem. Because most of our planet is covered by water, 70 to 90% of the human body (depending on age) is water. Testing water quality is simple, but as water can exist in different states or phases (liquid, solid, and gas), it can be made simpler. Advanced research [2] has been completed to understand water phases, resulting in the discovery of a new phase for water liquid. Water quality can be evaluated in each of the four phases. The focus of our research is the solid water phase known as frozen water crystal. We define "Frozen Water Crystal", a water crystal A microscopic crystal observed at the tip of a protrusion formed when liquid water is dropped in the form of drops onto a petri dish and frozen. The structure of the crystal is three-dimensional, and the crystal structure differs depending on the information possessed by the water. Crystals are formed when water changes to a solid-state, such as when it is frozen at −25 to −30 °C. Depending on the origin of the water and the formation process, crystals are divided into three main types: snow crystals, ice crystals, and water crystals. From the shape of the crystal, the purity level and the texture are clearly reflected, which then enables us to assess the quality of the water. Depending on the environmental conditions and the impact of the surrounding elements, the same water can give many different shapes. Each type of shape of crystals can be considered to be unique, without repetition. Many studies have been performed on classifying snow crystals and ice crystals [3–6]. In recent years, the application of deep learning on crystals became popular in the research world with the publication of 3D Crystal classification [7]. However, no real classification and understanding of water crystals based on deep learning have been completed until now.

This article innovation is the application of artificial intelligence for the first time on a dataset of photos of water crystals known as "the 5K EPP Dataset" collected with the collaboration of the I.H.M General Research Institute. This dataset is composed of high-resolution photos captured by a microscope camera under laboratory conditions. All the photos have been stored and managed by the I.H.M General Research Institute. A simple water crystal structure definition has been proposed in this research, which references snow crystals classification from [5] and the EPP project. This definition provides an easy understanding of the water structure with a 2D image.

With non-prohibited purposes, the water crystal dataset can support other researchers interested in water with their dataset [8,9]. These results assess the affection of human beings on water crystals initialization and the quality of water.

Deep learning has been widely applied in many research fields and achieved surprising results, especially in image processing. Convolutional neural networks (CNNs) are a special kind of neural network analysing high dimensional features dataset such as image, video, etc. CNNs were developed with image processing in mind, which makes them computationally more efficient when compared to other multi-layer back-propagation neural networks. CNNs can be used to automatically extract features from the dataset, which simplifies the next process. These features are not only useful for specific tasks but also can help in other related tasks. This opens a new era for research with reduced effort to achieve good results. With its well-understood architecture, CNNs are nowadays used widely in many areas, including image and structure classification. However, deep neural networks (DNNs) trained by conventional methods with small datasets commonly show worse performance than traditional machine learning methods [10]. The deeper network requires more data to train. This limitation prevents the wide application of deep learning in any field, in which collecting and assembling big datasets is a challenge. With the 5K EPP dataset, we face the same problem. We use data transformation techniques to enhance our dataset and the fine-tuning method is used to help train the model so that the model can learn better from a small dataset. Class weight is also used in this research to solve the imbalanced dataset problem. To build a classifier with deep learning, we split our work into 2 main steps: feature extractor and classification. We built a deep learning model to extract meaningful features from the EPP dataset, then used those to classify water crystal structures. We used 2 different techniques to extract features from the EPP dataset: convolutional auto-encoder and Fine-tuning. The extracted features are then stacked in convolution layers to make a classifier. The convolutional auto-encoder (CAE) [11] has been widely applied in dimension reduction and image noise reduction. Because the CAE model can keep the spatial information of the original image and extract information gently by using the convolution layer, it is considered to be one of the most state-of-the-art techniques in deep learning nowadays. Furthermore, it is an unsupervised method and can be used with less effort than a supervised one. Fine-tuning is a useful method for improving the performance of a neural network. It helps the researchers achieve higher performance with less effort.

In fine-tuning, a model trained on a given task is used for another similar task. This method reduced the training time and effort to extract meaningful features from the original input. ImageNet pre-trained models have been used for the fine-tuning method. This paper is organized as follows. Related works are in Section 2, the 5K EPP dataset is described in

Section 3, our dataset study approach and methods are in Section 4, Section 5 describes the experimental results, and the conclusions follow in Section 6.

2. Related Works

With a research focus to improve precipitation measurements and forecast for over 50 years, scientific studies of meteorology and weather include the study of snowflakes, ice crystals, and water crystals. Snowflake studies provide some of the most detailed evidence of climate change. It impacts atmospheric science. One of the first attempts to catalog snowflakes was made in the 1930s by Wilson Bentley who created a method of photographing snowflakes in 1931, using a microscope attached to a camera. The Bentley Snow Crystal Collection (https://snowflakebentley.com/ accessed on 15 October 2020) includes about 6125 items. A general classification of snow crystals $T_a - s$ diagram was proposed by Nakaya [3], which provides the most perfect classification from a physical point of view, with 7 categories. These categories include needles, columns, fern-like crystals developed in one plane, the combination of column and plane crystals, rimed crystals, and irregular crystals. The crystal images were collected from a slope of Mount Takachi, near the center of Hokkaido Island. Magono [4] published an improved version of Nakaya's classification, with the modification of and a supplement to Nakaya's classification of snow crystals. The results were obtained by laboratory experiments and from meteorological observation. The new classification provides temperature and humidity conditions, which can describe the meteorological differences in groups of asymmetric or modified types of snow crystals. It provides 80 categories, modified from Nakaya's categories and adding some new categories as well. Thirty thousand microscopic photographs of snow crystals taken by the Cloud Physics Group were used in their research.

Kikuchi and his team [5] proposed a new classification with 121 categories to classify snow crystals, ice crystals, and solid precipitation particles. They qualified their classification as "global scale" or "global" because their observations were performed from the middle latitudes (Japan) to polar regions. This classification consisted of three levels: general, intermediate, and elementary—which are composed of 8, 39, and 121 categories, respectively. Interestingly, this classification can be used not only for snow crystals but also for ice crystals. The deep learning method has been widely applied in many research fields, especially with image datasets. However, it faces the problem of working from a limited dataset. Fortunately, with the advent of image collection methods, a method to collect snowflake images was proposed: the Multi-Angle Snowflake Camera (MASC) [12]. It was developed to address the need for high-resolution multi-angle imaging of hydrometeors in freefall and has resulted in datasets comprising millions of images of falling snowflakes. Several studies have been published resulting from this development. A new method to automatically classify solid hydrometeors based on MASC images was presented by Praz et al. [13]. In this research, they proposed a regularized multinomial logistic regression (MLR) model to output the probabilistic information of MASC images. That probability is then weighed on the three stereoscopic views of the MASC to assign a unique label to each hydrometeor. The MLR model was trained using more than 3000 MASC images labeled by visual inspection. This model achieved very high performance with a 95% accuracy. Hicks et al. [6] published an automatic method to classify snowflakes, collected via Multi-Angle Snowflake Camera (MASC). The training dataset contains 1400 MASC images. They used a convolutional neural network and residual network which had been pre-trained with ImageNet as a back-bone for their model. Snowflakes were sorted by geometrics and divided into 6 distinct classes. Then, the degrees of rimming was decided by another training process, which has three distinct classes. Although the accuracy of this research is only 93.4%, it does provide a new way to classify snowflakes or nature structures automatically.

Another research with the MASC dataset was proposed by Leinonen et al. [14]. In this research, they aimed to classify large-scale MASC dataset by unsupervised learning method, using generative neural network (GAN) [15] and K-medoids [16]. With the features extracted from the discriminator part of the GAN model, they used the K-medoids algorithm to cluster all the images (data points) into 16 classes/categories. This method not only shows the hierarchical clustering groups but also requires no human intervention with such a large dataset. However, MASC images mainly show the crystal's degree of riming, but not the crystal's structure. This is because these images were taken during the falling progress of snowflakes.

In this research, we focus on building a new definition for water crystal classification based on previous studies and using deep learning to automatically classify them.

3. The 5K EPP Dataset

The water crystals have been provided by the Emoto Peace Project (EPP) at the I.H.M General Research Institute (Tokyo, Japan). Crystals were produced from water samples collected from many countries and sources, with the help of scientists all around the world. Water samples from each bottle are produced by the same procedure in [9]:

- From each bottle, a drop (approximately 0.5 mL) of water is placed into each of the 50 Petri dishes. So, there are 50 waterdrops from each bottle;
- Those dishes are then placed on a tray in a random position in a freezer maintained at −25 to −30 °C. The random placements helps to ensure that potential temperature differences within the freezer would be randomized among the dishes;
- The dishes are then removed from the freezer, and placed in a walk-in refrigerator (maintained at −5 °C). A water crystal photo is taken on the top of each resulting ice drop using a stereo optical microscope at either 100× or 200×, depending on the presence and size of a crystal.

Known as the **5K EPP dataset** [17], this dataset contains 5007 crystal photos in total. Because the 5K EPP dataset contains very high-resolution images (5472 × 3648 pixels) and water crystals only occupy a small part in the images, we needed to preprocess each image to remove the background. We used Otsu's method [18] to automatically define the border around the crystals. The minimum rectangular box that can cover each water crystal was chosen to crop the background. This helps reduce the image size while retaining the details in the object. Because the size of the water crystal in each image is different, we resized the cut-off images to the same size, to fit with the input of our machine.

The preprocessed dataset was then sorted into 13 categories. Based on the knowledge from the EPP Laboratory experts, we chose those categories that appeared most frequently in the 5K EPP dataset as our labels. We built a tree-like diagram in Figure 1 to demonstrate how we split the 5K EPP dataset into smaller categories. The branches of the tree correspond to the category in the definition. Finally, we obtained 13 branches corresponding to 13 categories. The details are given in Table 1. We chose the most typical images for each category and labeled them with the predefined definition. We split the 5K EPP dataset into the training set and test set with ratios of 80 and 20, respectively. The scikit-learn (https://scikit-learn.org/ accessed on 15 October 2020) library was used to split the dataset randomly and guarantee the balance in the dataset.

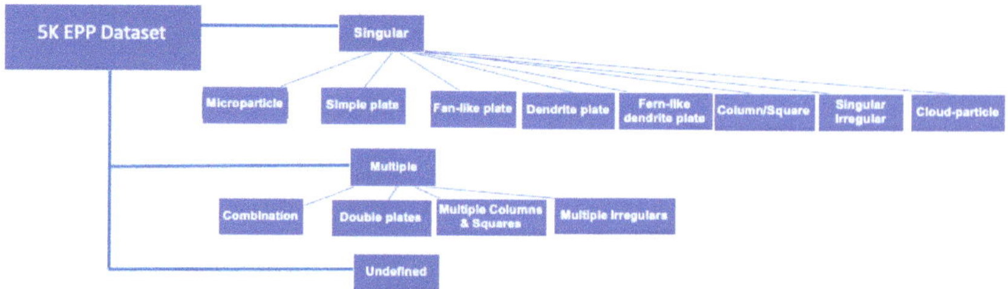

Figure 1. A tree-like diagram to demonstrate the water crystal categories with 5K EPP dataset.

Table 1. The definition for water crystal classes based on the knowledge from [5] classification.

Category	Crystal Example	Definition
Microparticule	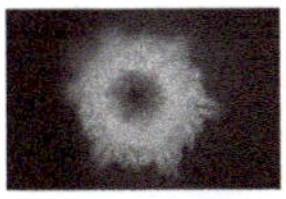	Crystal made up of fine particle on a hexagonal plate
Simple plate		Hexagonal crystal with no outer decoration
Fan-like plate	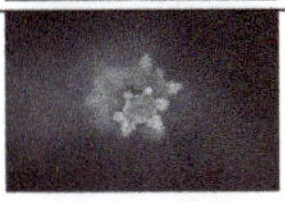	Square plate with a fan-shaped decoration on the outside
Dentrite plate		A square plate with dendritic decoration on the outside
Fern-like dendrite plate		A square plate with fern-like decorations on the outside
Column/Square	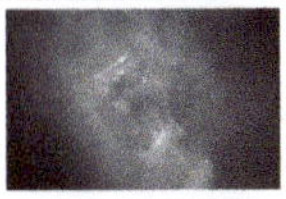	Square or columnar crystal/block crystal
Singular Irregular		Square plate with a fan-shaped decoration on the outside
Cloud-particle	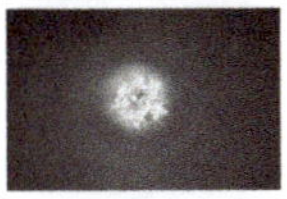	A granular decoration on a square plate
Combinations		Multiple square plates assembled together without overlapping vertically
Double plate		Two square plates stacked on top of each other
Multiple Columns/ Squares		Multiple square or columnar crystals / Multiple block crystals
Multiple Irregulars	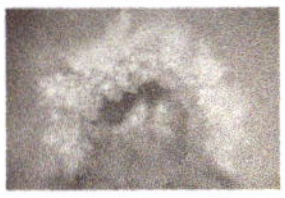	Multiple asymmetrical crystals or crystals that are not fully formed
undefined	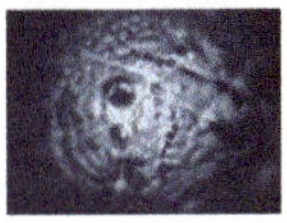	Types of water crystals without crystals

4. Proposed Method

4.1. Feature Extractor

4.1.1. Residual Auto-Encoder

A convolutional auto-encoder (CAE) is an efficient technique used to reduce dimensionality and generate high-level representation from raw data. It is an unsupervised learning algorithm using a back propagation algorithm to update parameters. In this model, the targets are equal to the inputs. A convolutional auto-encoder is composed of two models: an encoder and a decoder. The encoder aims to find the latent representation for input data, while the decoder is tuned to reconstruct the original input from the encoder's output.

Considering a dataset X with n sample and m features, the encoder learns the latent representation H and the decoder tries to reconstruct the original input X' from H, by minimizing the differences between X and X' over all samples:

$$\min_{W,W'} \frac{1}{n} \sum_{i=1}^{n} \|D_{W'}(E_W(X_i)) - X_i\|_2^2.$$

For a convolutional auto-encoder,

$$E_W(X) = \sigma(X * W) = H$$

$$D_{W'}(H) = \sigma(H * W') = X'$$

where W and W' are learnable parameters and σ is the activation function such as ReLU and sigmoid. At the end of the training process, the embedded code H is used as a new representation of input X. Then, H can be fed into a fully connected layer to do classifying or clustering tasks. We proposed a new CAE model to extract latent representation from high-resolution water crystal images. First, in the encoder, 3 convolution layers were stacked on the input images to extract latent representation. Then, the encoder's output was flattened to form a vector, which is an extracted feature. The decoder transformed embedded features back to the original image. The convolution (transpose) layers with stride allow the network to learn spatial subsampling (upsampling) from data, leading to a higher capability of transformation. Therefore, instead of using a convolution layer followed by a pooling layer, we used a convolution layer with a stride in the encoder and a convolution transpose layer with a stride in the decoder. To achieve low-dimension images with high-representation with very high-resolution images was a challenging task. Down-sampling images to get low dimension representation can lead to a vanishing gradient when training a very deep neural network model. With a traditional CAE, the greater number of hidden layers, the hard it is to reconstruct the original image. To solve this problem, we used the skip idea from ResNet [19]: skip connection. Skip connection addresses the problem with vanishing gradient and information lossless. The idea is that instead of letting the model learn underlying mapping, let it learn the residual mapping.

With skip connection, the residual or identity was added to the output. We obtained the output defined as follows: $y = R(x) + x$.

Because we had the identity connection come due to x, the model actually learned the residual $R(x)$. We used two different kinds of residual block to build an encoder block: a regular block and a downsample block. As the regular block, the residual block has 3 convolution layers with the same number of output channels. The downsample block decreases the sampling rate of the input by deleting samples. When the block performs frame-based processing, it resamples the data in each column of the Mi-by-N input matrix independently. When the block performs sample-based processing, it treats each element of the input as a separate channel and resamples each channel of the input array across time. The resample rate is K times lower than the input sample rate, where K is the value of the Downsample factor parameter. The Downsample block resamples the input by discarding K – 1 consecutive samples following each sample that is output.

Each convolution layer was followed by a batch normalization layer and a ReLU activation function. Then, we skipped these three convolution operations and added the input directly before the final ReLU activation function. This kind of design requires that the output of the three convolution layers be of the same shape as the input, so they can be added together. The downsample block had the same design as the regular one, but the first convolution layer reduced the image size and had a different number of channels. To add the input before the last ReLU activation function, we used a 1 × 1 convolution layer, followed by a batch normalization layer, to transform the input into the desired shape for the addition operation. By experimentation, we found that using two convolution layer after the first convolution layer in each block helps the model reconstruct output better.

The first convolution layer has been used to downsize the image by two and the following two were used to learn useful information. Each convolution layer is followed by a batch normalization layer and an activation layer (except the last one). In this research, we chose ReLU as an activation layer. The skip connection used convolution and batch normalization to reduce the size of the input so that it was equal to the output. The final architecture is shown in Figure 2. We used one convolution layer and three residual blocks to build an encoder. The decoder kept the same structure as the origin. The final model is called a residual auto-encoder (RAE). The reconstruction loss was used to evaluate the performance of the RAE model. The parameters of encoder and decoder were updated by minimizing the reconstruction error:

$$L_r = \frac{1}{n}\sum_{i=1}^{n} Distance(D_{W'}(E_W(x_i)), x_i).$$

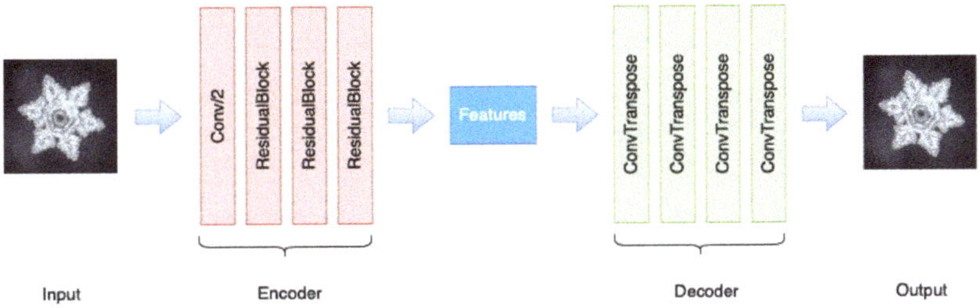

Figure 2. A residual auto-encoder model to extract features from origin images. Each residual block is a combination of a Downsample block and a regular block, respectively.

Instead of using the Euclidean distance to compute the reconstruction error, we used the Spherical distance in [20]. The latent representations extracted from the RAE model were projected into the surface of the unit hyper-sphere. The distance between data points in that surface was then measured by $d_{spherical}$ function, which is defined as follows:

$$d_{spherical} = \frac{\arccos(s_{\cos(z_i,z_j)})}{\pi} = \frac{1}{\pi}\langle \frac{z_i}{||z_i||_2+\epsilon}, \frac{z_j}{||z_j||_2+\epsilon}\rangle$$

where $\arccos(\alpha)$ is the inverse cosine function for $\alpha \in [-1,1]$ and ϵ is a very small value to avoid numerical problem.

4.1.2. Fine-Tuning Model

The efficiency of the classification model is based on the power of the features extracted from the training dataset. With the high meaning features, the classifier can achieve good results from the very first training steps. Auto-encoder is a popular strategy used to extract features from the unlabeled dataset. Because it requires no label to train the CNN model, it can perform well on high dimension datasets, especially images and videos. The model then trains with our full dataset and learns the most important information from these images. However, to choose a good architecture for the auto-encoder and train it from scratch is not easy as it requires a great deal of knowledge about machine learning and specific datasets. A new method was introduced to help to solve problems with feature extracting, called fine-tuning. Fine-tuning is a process that takes a network that has already been trained for a given task and make it perform a second similar task. Many studies have been shown that fine-tuning techniques can get good results with less effort compared to starting from scratch.

For image related tasks, the most common way is fine-tuning the model trained on ImageNet [21] (with 1.2 million labeled images) by continuing to train it on the original dataset. A competition on classification and object detection has been organized to find state-of-the-art techniques to solve those problems on the ImageNet dataset, called ImageNet Large Scale Visual Recognition Challenge (ILSVRC) (http://www.image-net.org/challenges/LSVRC/ accessed on 15 October 2020). AlexNet [22] is the first larger-scale convolution model that does well on the ImageNet classification task, outperforming all previous non-deep learning-based models by a significant margin, and won the competition in 2012. After that, VGG [23] was proposed with the idea of deeper networks and much smaller filters, which is a significant jump in deep learning. ResNet [19] introduced residual blocks with skip connections, which allow the gradients to backpropagate through to the initial layers without vanishing. That model won the first prize in the ILSVRC 2015 competition with an error rate of just 3.6%. In 2016, a new model was proposed, called SqueezeNet [24]. This model achieved approximately the same level of accuracy as AlexNet with a much smaller number of parameters. So, it is suitable for building mobile applications. In the same year, DenseNet [25], a densely connected network, was proposed to improve higher layer architectures of the previously developed networks. The DenseNet architecture attempts to solve this problem by densely connecting all the layers: each layer gets the input from the previous layer's output. We used all those models as back-bone to build a model to classify our water crystals. All the experiment results are shown in Table 3.

4.2. Classification Model

The features extracted from each previous step are then feed into the classifier layers to build the classification model.

The classifier has 2 main parts: feature extractor and classifier. To build the extractor, RAE pre-trained and Image pre-trained models were used. With the RAE model, we keep only the encoder from the RAE model, which had been pre-trained with the EPP dataset, as a feature extractor. As with the ImageNet pre-trained models, the last layer is removed to get the latest features. The classifier includes three fully connected layers which are added on top of the feature extractor and then trained simultaneously with the labeled dataset. The overview of the final classification model is given in Figure 3.

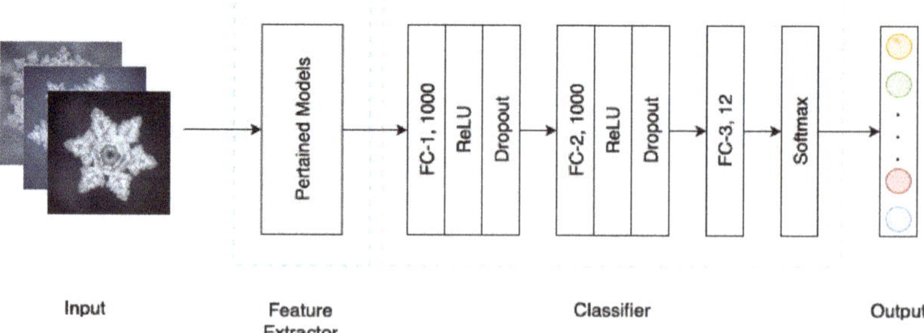

Figure 3. Classification architecture overview. The feature extractor can be replaced by RAE pre-trained model or ImageNet pre-trained model. The features are used as input for the next step. The classifier contains 3 fully connected layers, each of them is followed by ReLU and a dropout layer. The last FC outputs the predicted probability distribution. Softmax is added to get the final prediction.

Unlike training a model from scratch, we unfreeze early layers and train the whole network. A small learning rate has been chosen to let the classifier learn the patterns from the previously learned convolution layers in the pre-trained network. For further evaluation and improvement, we chose different metrics to compare the performance between different feature extractors. The comparison results are then shown in Section 5.

4.3. Imbalanced Data

Due to the crystal formation process in nature, the amount of data in each class is imbalanced. Therefore, when labeling the 5K EPP dataset, we realized that there is an imbalance between the number of images among the categories. Some categories include approximately 20% dataset while some others include just 2%. The details are provided in Table 2.

Table 2. The 5K EPP dataset summary.

Category	Card(Photo)	Percentage
Microparticle	161	3.2%
Simple plate	104	2%
Fan-like plate	341	6.81%
Dendrite plate	1388	27.72%
Fern-like dendrite plate	674	13.46%
Column/Square	38	7.5%
Singular Irregular	674	13.46%
Cloud-particle	3	0.0006%
Combination	129	2.57%
Double plates	204	4%
Multiple Columns/Squares	172	3.4%
Multiple Irregular	692	13.82%
Undefined	427	8.52%

To guarantee balance and accuracy when training the deep learning model, we used the class weight method. We simply provided a weight for each class which places more emphasis on the minority classes. Following that idea, the model can learn from all classes equally. Each class will be assigned a weight corresponding to the number of images inside. The weight can be calculated as follows:

$$w_i = \frac{N}{C * n_i}$$

where w_i, n_i, C, and N are the weight assigned to class i, the number of images of class i, the number of classes, and the total images of dataset, respectively. We also use F_1-score metric to evaluate the model performance with an imbalanced dataset besides the standard evaluation metric. Both are described in Section 5.1.

5. Experiments and Results

5.1. Evaluation Metric

5.1.1. Classification Accuracy

We used standard evaluation metrics to evaluate classification results. For all implementation setup, we set the number of classes equal to the number of ground-truth categories that were used to label the dataset in Section 3. The performance is evaluated by the accuracy metric:

$$ACC = \frac{1}{n}\sum_{i=1}^{n}(y_i^{true} = y_i^{pred}) \quad (1)$$

where y^{true} is the ground-truth label, y^{pred} is the prediction label, and n is number of images inside the test set. The test dataset is not used when training the model. The best model should have high accuracy for both training and test progress.

5.1.2. F_1-Score

With the imbalanced dataset, an efficient way to evaluate the model performance is using F_1-score [26]. Instead of calculating the ratio of true prediction within the total images, F_1-score measures accuracy by precision p and recall r. The formula for F_1-score is defined as follows:

$$F_1 = 2 * \frac{precision * recall}{precision + recall} \quad (2)$$

When p is the number of correct positive results divided by the number of all positive results returned by the classifier, and r is the number of correct positive results divided by the number of all relevant samples (all samples that should have been identified as positive).

5.2. Experiments Environment and Setup

In this section, we discuss applying different pre-trained models used in fine-tuning with our 5K EPP dataset. For our experiments, we used an NVIDIA Tesla V100 SXM2 GPU, with 32GB of memory. The server used for running the experiments was Grid5000 [27] (https://www.grid5000.fr accessed on 15 October 2020), the French large-scale and flexible experimental grid platform consisting of 8 sites geographically distributed over France and Luxembourg. Each site comprises one or several clusters, for a total of 35 clusters inside Grid5000. This platform is dedicated to experiment-driven researches in all areas of computer science, with a focus on parallel and distributed computing including Cloud, HPC, and Big Data and AI. Our implementation is based on Python and Pytorch (https://pytorch.org accessed on 15 October 2020).

5.3. Experiment Results

5.3.1. Residual Auto-Encoder Model (RAE)

We first trained the RAE model with an unlabelled dataset. Adam optimizer [28] was used to update model parameters, with learning rate $\alpha = 10^{-4}$. Regularization was used to reduce the overfitting problem, with $\gamma = 10^{-5}$. We chose the number of images per batch equal to 32. The model was trained with 100 epochs. We used two different loss functions to train the RAE model: one Spherical citetran2019deep metric and the Binary Cross-Entropy (noted BCE). The reconstruct results built with both metrics are shown in Figure 4. Although the BCE loss function can reconstruct an image quite similar to the original one, when zooming out the image, we can see that some parts of the image are blurred and old content cannot be seen. With Spherical, the reconstructed image is the same as the original one. We also used the Structural Similarity Index (SSIM) [29] to assert the similarity among reconstructed images and the input. The average SSIM has been computed for both Spherical and BCE. In overall, Spherical's SSIM is 0.96, while the BCE's SSIM is 0.89. Therefore, we concluded that Spherical outperformed BCE.

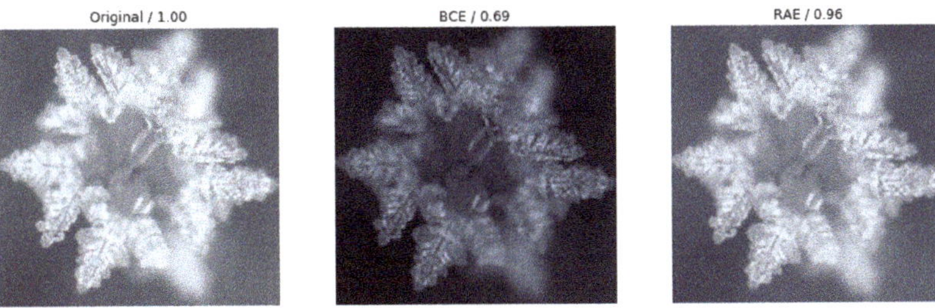

Figure 4. Reconstruct image generated by RAE model train with BCE and Spherical metric separately. The SSIM index is calculated with each reconstructed image. Spherical one outperforms the BCE one.

5.3.2. Classification Model

We trained the classification models proposed in Section 4. Stochastic Gradient Descent (SGD) with Nesterov momentum optimizer was used to update parameters, with learning rate $\alpha = 10^{-3}$, momentum $\Delta = 0.9$ and regularization $\gamma = 10^{-4}$. To enrich the dataset, we used transform techniques such as flip image (vertical and horizontal) with probability $p = 0.5$, rotating the image with a random degree in the range from −90 to 90 degree, random cropping. The classification model is first trained with 100 epochs.

As in Section 4, we used 2 kinds of pre-trained models to build classifiers: RAE pre-trained model and ImageNet pre-trained models. RAE model was trained with an unlabeled dataset, as mentioned in the previous result. With ImageNet pre-trained models, we chose the most popular deep learning models, which had won the ILSVRC: AlexNet, VGG, SqueezeNet, DenseNet, and ResNet. All parameters were adjusted during the training progress. The 5K EPP dataset was then divided into a training and a test set with a ratio of 80:20.

After visualizing and doing statistics on the prediction, we realized that the definition with 13 categories for water crystal was not too good and gave unclear instructions. The classification model sometimes misclassified between water crystal and its spatial form. Because we used a 2D image to classify the plates, it is hard to see the differences between a plate with and without space elements. As mentioned in [30], we should apply machine learning in a problem that humans can do well. We modified the definition to eliminate ambiguity in the labeling process. The major change was combining the space misclassified categories. Additionally, we also found a new category named double plates. Finally, we delivered 12 categories, which are defined in Section 3.

Two fully-connected layers are added on the top of modified models, followed by a ReLU activation layer. In addition to regularization, we also used the traditional dropout method to prevent overfitting problems [31]: a dropout layer is put after each fully-connected layer, except the last one. We fine-tuned parameters and the applied data transformation techniques mentioned in Section 3 to enrich the dataset.

With the new dataset, the model has kept training with the same configuration and parameters. The new definition obtained significant performance improvement. The model could overcome the overfitting problem and obtain high accuracy for both training and test progress.

We used different pre-trained models as back-bone and trained the model with the same parameters and set up conditions. The standard accuracy and F_1-score were calculated and compare among models. The results are shown in Table 3. The model trained with RAE outperforms models that used AlexNet and SqueezeNet as back-bone, with 4% higher than AlexNet and 8% higher than SqueezeNet in F_1-score metric. Although other pre-trained models (such as VGG, DenseNet) have high accuracy, the F_1-scores of Alexnet and SqueezeNet are much lower. In addition, the loss values of VGG and DensetNet are two times bigger than the lowest one (e.g., RestNet). ResNet outperforms other models in both loss value and accuracy with 98.50% Top-1 accuracy and 97.25% F_1-score. We concluded that the ResNet back-bone is the best solution for our problem. When evaluating the model with test set, the ResNet accuracy is approximately 93%.

Table 3. Top-1 Accuracy and F_1-score on 5K EPP training set.

Backbone	Loss	Accuracy	F_1-Score
RAE	0.094	94.35%	91.64%
AlexNet	0.086	93.71%	87.79%
VGG	0.049	96.21%	92.03%
SqueezeNet	0.130	91.16%	83.31%
DenseNet	0.046	96.93%	93.55%
ResNet	0.025	98.50%	97.25%

5.3.3. Comparative Model

To demonstrate the effectiveness of our model, we compared it with Hicks's model [6]. Both methods used ResNet pre-trained models as the backbone and used the fine-tuning method to train all the parameters.

In their study, Hicks et al. implemented a classifier to automatically classify geometrically and riming-degree of the MASC dataset. They used the ResNet pre-trained model to initialize the model parameter and added a new FC layer as a classifier layer. The model outputs the probability of 6 distinct snowflake categories, which is defined by Hicks et al. They used 2 CNN models to do distinct tasks: (1) classify crystal geometrics and (2) classify riming-degree. Based on the crystal structure classification purpose, we compared our model with Hicks' first model. We trained Hicks' model with the 5K EPP dataset and used classification accuracy to compare its performance to ours. The results are shown in Figure 5. Even though our accuracy is just slightly higher than Hicks', the training progress can show that our model is more stable and the convergence of our model is better than that of the Hicks'model.

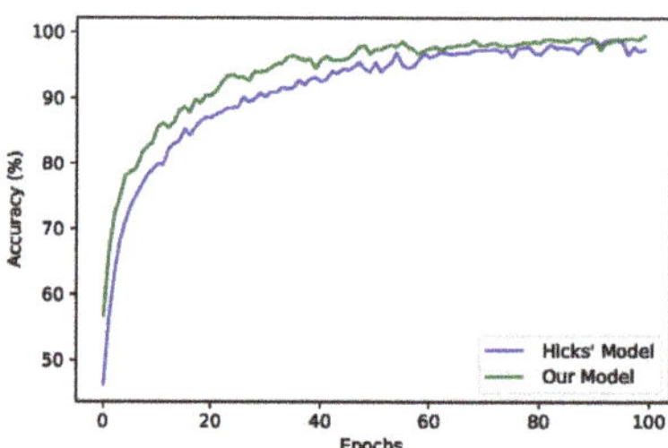

Figure 5. Our proposed model compared with Hicks's model. Both implementations are trained on the 5K EPP dataset.

6. Conclusions

Based on the EPP water crystal dataset and the previous knowledge about snowflake classification, we proposed a simple water crystal definition, which can be used to classify the EPP dataset. We contributed a new data science dataset, called the 5K EPP dataset, with 5007 images split into 13 classes (12 categories + undefined). We proposed a deep learning-based method to automatically classify this dataset. We compared fine-tuning results between the residual auto-encoder model, trained with unlabelled EPP datasets, and ImageNet pre-trained models, and then selected the best one. With a fine-tuning technique and ResNet pre-trained model, we had a classifier with 93% accuracy. With this result, we are going to extend the 5K EPP dataset by applied the water crystal definition to label the EPP water crystal 20K dataset. A new approach to using an unsupervised method to deal with the unlabeled dataset and find a new group of the water crystal structure will be targeted in further studies.

Author Contributions: Conceptualization, H.D.T., F.A. and L.T.Q.; methodology, H.D.T. and L.T.Q.; software, H.D.T.; validation, H.D.T., F.A., L.T.Q., H.E., M.H., K.K. and T.O.; formal analysis, H.D.T. and L.T.Q.; investigation, H.D.T., F.A., K.K., and T.O.; resources, K.K. and T.O.; data curation, K.K., T.O., M.H., F.A. and H.D.T.; writing—original draft preparation, H.D.T., F.A. and L.T.Q.; writing—review and editing, H.D.T., F.A., L.T.Q., H.E., M.H. and K.K.; visualization, H.D.T., F.A., L.T.Q., H.E., M.H. and K.K.; supervision, F.A., L.T.Q., H.E., M.H., K.K. and T.O.; funding acquisition, F.A. All authors have read and agreed to the published version of the manuscript.

Funding: This work was supported by the National Institute of Informatics (NII) under the GLO Internship program and Emoto Peace Project, Non-profit Organization.

Informed Consent Statement: Not applicable.

Data Availability Statement: Data will be available for the data science.

Acknowledgments: We would like to thank you the National Institute of Informatics (Tokyo, Japan) for the support of the research and I.H.M General Research Institute (Tokyo, Japan) for their help in the Water Crystal classification. We are grateful to D'Orazio and the French Grid5000 programs (https://www.grid5000.fr (accessed on 15 October 2020)) for providing the grid infrastructures, advice, and user assistance.

Conflicts of Interest: The authors declare no conflict of interest.

References

1. Boyd, C.E. *Water Quality: An Introduction*, 3rd ed.; Springer Nature Switzerland AG: Berlin/Heidelberg, Germany, 2020. [CrossRef]
2. Pollack, G. *The Fourth Phase of Water: Beyond Solid, Liquid and Vapor*; Ebner & Sons: Springfield, OH, USA, 2013.
3. Nakaya, U. *Snow Crystals: Natural and Artificial*; Hokkaido University: Hokkaido, Japan, 1954.
4. Magono, C.; Lee, C.W. Meteorological classification of natural snow crystals. *J. Fac. Sci. Hokkaido Univ. Ser. 7 Geophys.* **1966**, *2*, 321–335.
5. Kikuchi, K.; Kameda, T.; Higuchi, K.; Yamashita, A.; Working Group Members for New Classification of Snow Crystals. A global classification of snow crystals, ice crystals, and solid precipitation based on observations from middle latitudes to polar regions. *Atmos. Res.* **2013**, *132*, 460–472. [CrossRef]
6. Hicks, A.; Notaroš, B. Method for Classification of Snowflakes Based on Images by a Multi-Angle Snowflake Camera Using Convolutional Neural Networks. *J. Atmos. Ocean. Technol.* **2019**, *36*, 2267–2282. [CrossRef]
7. Ziletti, A.; Kumar, D.; Scheffler, M.; Ghiringhelli, L.M. Insightful classification of crystal structures using deep learning. *Nat. Commun.* **2018**, *9*, 2775. [CrossRef] [PubMed]
8. Radin, D.; Hayssen, G.; Emoto, M.; Kizu, T. Double-blind test of the effects of distant intention on water crystal formation. *Explore* **2006**, *2*, 408–411. [CrossRef] [PubMed]
9. Radin, D.; Lund, N.; Emoto, M.; Kizu, T. Effects of distant intention on water crystal formation: A triple-blind replication. *J. Sci. Explor.* **2008**, *22*, 481–493.
10. Feng, S.; Zhou, H.; Dong, H. Using deep neural network with small dataset to predict material defects. *Mater. Des.* **2019**, *162*, 300–310. [CrossRef]
11. Masci, J.; Meier, U.; Cireşan, D.; Schmidhuber, J. Stacked convolutional auto-encoders for hierarchical feature extraction. In *International Conference on Artificial Neural Networks*; Springer: Berlin/Heidelberg, Germany, 2011; pp. 52–59.
12. Garrett, T.; Fallgatter, C.; Shkurko, K.; Howlett, D. Fall speed measurement and high-resolution multi-angle photography of hydrometeors in free fall. *Atmos. Meas. Tech.* **2012**, *5*, 2625–2633. [CrossRef]
13. Praz, C.; Roulet, Y.A.; Berne, A. Solid hydrometeor classification and riming degree estimation from pictures collected with a Multi-Angle Snowflake Camera. *Atmos. Meas. Tech.* **2017**, *10*, 1335–1357. [CrossRef]
14. Leinonen, J.; Berne, A. Unsupervised classification of snowflake images using a generative adversarial network and K-medoids classification. *Atmos. Meas. Tech.* **2020**, *13*, 2949–2964. [CrossRef]
15. Goodfellow, I.; Pouget-Abadie, J.; Mirza, M.; Xu, B.; Warde-Farley, D.; Ozair, S.; Courville, A.; Bengio, Y. Generative adversarial nets. In *Advances in Neural Information Processing Systems*; MIT Press: Cambridge, MA, USA, 2014; pp. 2672–2680.
16. Jain, A.K.; Murty, M.N.; Flynn, P.J. Data clustering: A review. *ACM Comput. Surv. (CSUR)* **1999**, *31*, 264–323. [CrossRef]
17. Emoto, H.; Doan Thi, H.; Andres, F.; Hayashi, M.; Katsumata, K.; Oshide, T.; Tran, L. 5K EPP Dataset 2021. Available online: https://ieee-dataport.org/documents/5k-epp-dataset (accessed on 15 October 2019). [CrossRef]
18. Otsu, N. A threshold selection method from gray-level histograms. *IEEE Trans. Syst. Man Cybern.* **1979**, *9*, 62–66. [CrossRef]

19. He, K.; Zhang, X.; Ren, S.; Sun, J. Deep residual learning for image recognition. In Proceedings of the IEEE Conference on Computer Vision and Pattern Recognition, Las Vegas, NV, USA, 27–30 June 2016; pp. 770–778.
20. Tran, B.; Le Thi, H.A. Deep Clustering with Spherical Distance in Latent Space. In *International Conference on Computer Science, Applied Mathematics and Applications*; Springer: Berlin/Heidelberg, Germany, 2019; pp. 231–242.
21. Deng, J.; Dong, W.; Socher, R.; Li, L.J.; Li, K.; Fei-Fei, L. Imagenet: A large-scale hierarchical image database. In Proceedings of the 2009 IEEE Conference on Computer Vision and Pattern Recognition, Miami, FL, USA, 20–25 June 2009; pp. 248–255.
22. Krizhevsky, A.; Sutskever, I.; Hinton, G.E. Imagenet classification with deep convolutional neural networks. In *Advances in Neural Information Processing Systems*; MIT Press: Cambridge, MA, USA, 2012; pp. 1097–1105.
23. Simonyan, K.; Zisserman, A. Very deep convolutional networks for large-scale image recognition. *arXiv* **2014**, arXiv:1409.1556.
24. Iandola, F.N.; Han, S.; Moskewicz, M.W.; Ashraf, K.; Dally, W.J.; Keutzer, K. SqueezeNet: AlexNet-level accuracy with 50x fewer parameters and <0.5 MB model size. *arXiv* **2016**, arXiv:1602.07360.
25. Huang, G.; Liu, Z.; Van Der Maaten, L.; Weinberger, K.Q. Densely connected convolutional networks. In Proceedings of the IEEE Conference on Computer Vision and Pattern Recognition, Honolulu, HI, USA, 21–26 July 2017; pp. 4700–4708.
26. Goutte, C.; Gaussier, E. A probabilistic interpretation of precision, recall and F-score, with implication for evaluation. In *European Conference on Information Retrieval*; Springer: Berlin/Heidelberg, Germany, 2005; pp. 345–359.
27. Balouek, D.; Carpen Amarie, A.; Charrier, G.; Desprez, F.; Jeannot, E.; Jeanvoine, E.; Lèbre, A.; Margery, D.; Niclausse, N.; Nussbaum, L.; et al. Adding Virtualization Capabilities to the Grid'5000 Testbed. In *Cloud Computing and Services Science*; Ivanov, I.I., van Sinderen, M., Leymann, F., Shan, T., Eds.; Communications in Computer and Information Science; Springer International Publishing: Berlin/Heidelberg, Germany, 2013; Volume 367, pp. 3–20. [CrossRef]
28. Kingma, D.P.; Ba, J. Adam: A method for stochastic optimization. *arXiv* **2014**, arXiv:1412.6980.
29. Wang, Z.; Bovik, A.C.; Sheikh, H.R.; Simoncelli, E.P. Image quality assessment: From error visibility to structural similarity. *IEEE Trans. Image Process.* **2004**, *13*, 600–612. [CrossRef]
30. Ng, A. Machine Learning Yearning. 2017. Available online: http://www.mlyearning.org/(96) (accessed on 15 October 2019).
31. Srivastava, N.; Hinton, G.; Krizhevsky, A.; Sutskever, I.; Salakhutdinov, R. Dropout: A simple way to prevent neural networks from overfitting. *J. Mach. Learn. Res.* **2014**, *15*, 1929–1958.

Article

Combination of Transfer Learning Methods for Kidney Glomeruli Image Classification

Hsi-Chieh Lee * and Ahmad Fauzan Aqil

Department of Computer Science and Information Engineering, National Quemoy University, Kinmen 89250, Taiwan; afauzanaqil@gmail.com
* Correspondence: cjlee@email.nqu.edu.tw; Tel.: +886-932066603

Abstract: The rising global incidence of chronic kidney disease necessitates the development of image categorization of renal glomeruli. COVID-19 has been shown to enter the glomerulus, a tissue structure in the kidney. This study observes the differences between focal-segmental, normal and sclerotic renal glomerular tissue diseases. The splitting and combining of allied and multivariate models was accomplished utilizing a combined technique using existing models. In this study, model combinations are created by using a high-accuracy accuracy-based model to improve other models. This research exhibits excellent accuracy and consistent classification results on the ResNet101V2 combination using a mix of transfer learning methods, with the combined model on ResNet101V2 showing an accuracy of up to 97 percent with an F1-score of 0.97, compared to other models. However, this study discovered that the anticipated time required was higher than the model employed in general, which was mitigated by the usage of high-performance computing in this study.

Keywords: combined classification model; deep transfer learning; focal-segmental; kidney disease; kidney glomeruli; medical image; sclerosed glomeruli

Citation: Lee, H.-C.; Aqil, A.F. Combination of Transfer Learning Methods for Kidney Glomeruli Image Classification. *Appl. Sci.* **2022**, *12*, 1040. https://doi.org/10.3390/app12031040

Academic Editor: Giancarlo Mauri

Received: 31 August 2021
Accepted: 17 January 2022
Published: 20 January 2022

Publisher's Note: MDPI stays neutral with regard to jurisdictional claims in published maps and institutional affiliations.

Copyright: © 2022 by the authors. Licensee MDPI, Basel, Switzerland. This article is an open access article distributed under the terms and conditions of the Creative Commons Attribution (CC BY) license (https://creativecommons.org/licenses/by/4.0/).

1. Introduction

Acute kidney injury is a common and significant complication. And it has been proven that one of the factors that cause the disease to occur frequently is the increase in the widespread distribution of red blood cells and excessive inflammation. [1,2]. The pathophysiological mechanisms that explain the association between increased RDW scores and poorer prognosis remain unclear. According to current knowledge, an increase in RDW is the cause of microcirculation disorders. Older erythrocytes gradually lose the ability to damage cell membranes. This feature is especially important during the squeezing of nucleated cells through small diameter vessels in organs, such as the kidneys. The stiff and large erythrocytes observed in patients with high values of RDW could not enter through the capillaries and thus impaired blood flow through the microcirculation, leading to ischemia of the renal tissue [1].

Recently, kidney disease is increasingly being found in the spread of the coronavirus during the current pandemic [3]. The identification of the host of the second angiotensin-converting enzyme (ACE2) is the first step in the entry of viral infections into the body, where this host leads to cell fusion and entry into host cells in the lungs. Other epithelial cells, such as renal cells, have a high ACE2 expression [4]. As a result, several studies have shown that the long-term consequences of COVID-19 infection may cause COVID-19 patients to develop chronic kidney disease. The antigen capture technique showed that an increase in the SARS-CoV-2 protein in urine samples was detectable when compared to patients before the epidemic. By considering this condition, kidney disease detection needs some prerequisites to obtain the results quickly and accurately. Since COVID-19 is implicated in kidney function, this study aims in identifing how the anatomy of kidney disease is affected by COVID-19. Histopathology revealed that abnormalities in the kidney occur when the shape of the glomerulus in the kidney structure is different, eventually leading to chronic kidney disease [5]. Understanding how the complex structure of kidney tissue differs, requires the use of imaging technology to compare normal and diseased kidneys.

Until recently, there were two types of glomerular tissue disorders: sclerosed and focal-segmental sclerosed [6]. The difference between these two types may be seen in Bowman's capsule structure around the glomerular network. Bowman's visible capsule surrounds the glomerulus, suggesting that the glomerular tissue is undeniably normal, whereas Bowman's capsule that seems faint or even disappears indicates that the glomerular tissue is unhealthy [7]. This explanation is supported by the morphological and Karpinski scores, which indicated the presence of a difference between nucleus and capillary lumens, and the number of areas (mesangial matrix) on the typical glomerular capillary lumens were absent. Moreover, the Bowman's capsule was filled by collagen in glomerular non-healthy kidneys. Focal segmental sclerosis is a disease of the glomerulus that affects many people. A difference in glomerular size, the degree of leg process elimination and the alteration do the celial cells are all signs of this illness [6]. Figure 1 depicts the anatomy distinctions among three types of glomeruli. Imaging technology is predicted to aid in the accurate detection of individuals with renal disease, allowing for better medical performance. Furthermore, the model's ability to analyze data will be determined by how much time is allotted to it. The transfer learning technique used in this study was taken from several machine learning models after reviewing various methods by considering the procedures and work results obtained. Bueno et al. [8] used UNet and SegNet to divide the glomerulus into three groups based on segmentation pixels. The data processing findings demonstrate that data prediction from the train data is accurate to 98.16 percent.

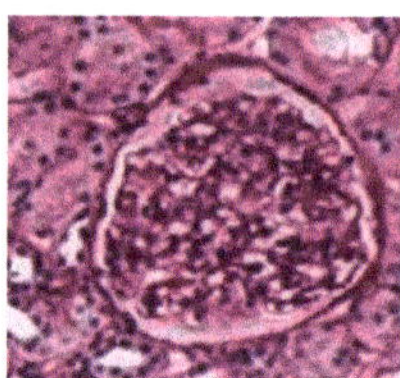
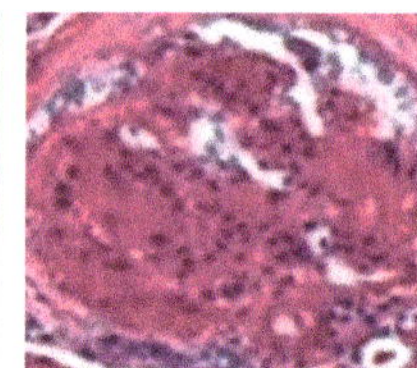
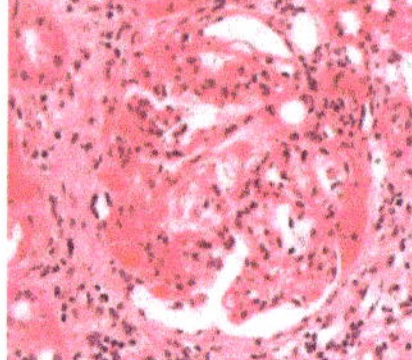

Figure 1. Anatomies of normal glomeruli, sclerosed glomeruli and focal-segmental glomeruli. (**Left**) Normal glomeruli; (**Center**) sclerosed glomeruli and (**Right**) focal-segmental glomeruli.

As a result, the combination of these approaches is used as a benchmark for comparing the proposed study with the EfficientNet method [9]. In the previous two years, this technique has been enhanced by altering the model's structure, combining the models and employing the iterative model process. The ImageNet dataset, which comprises random pixel data (non-medical picture), has recently been used in some studies to show that this method's accuracy rate approaches 99.70 percent [10]. This result is utilized as a proof-of-concept that the EfficientNet technique can provide high accuracy when utilizing the medical picture dataset. Therefore, the goal of this research is to demonstrate how this approach may be used in medical imaging collections.

The majority of deep learning in the categorization of glomeruli comprises of normal, sclerosed and non-glomerular classes, according to the prior studies. As a result, this study proposes three possible classifications: normal, sclerosed and focal-segmental sclerosed. Since focal-segmental sclerosed glomeruli [11] are a type of sclerosis that cause anomalies in the glomerular tissue and affect a large number of people, it is critical to correctly detect the focal-segmental anatomy of sclerosed glomeruli in kidney disease diagnosis.

The normal glomerulus, sclerosed glomerulus and focal-segmental glomerulus were all used to create this approach for transfer learning. Sections 2 and 3 provide detailed explanations of the database, materials and research techniques, whereas Sections 4 and 5 contain the results of the tests and conclusions.

2. Materials and Methods

This section details the procedures involved in conducting the study, including the data sources, methodologies and models used to enhance and manage the studies for improved outcomes.

2.1. Data and Preparation

The experimental data consisted of 5095 biomolecular pictures of the kidneys in png format, each of which was derived from 2926 photos of the segmentation data conducted by Bueno et al. [8,12]. These data can be used to benchmark the assessment data in a test classification system that divides the data into normal glomeruli and sclerosed glomeruli. Dimitris, in the Kaggle dataset [13], separated the 1584 pictures from the open data challenges, according to the form of the glomeruli tiles, into raw data and 585 additional focal-segmental images obtained from the study obtained by Kannan et al. [14].

The picture has a resolution of 256 × 256 pixels and is divided into three categories: normal, sclerosed, and focal-segmental. This clustering was performed to show the part of the experimental section's categorization findings that were correct. Rosenberg et al. [6] conducted studies on focal-segmental sclerosis. Martin-Navarro [11] discovered focal-segmental kidney disease types in pulmonary sarcoidosis, while Asinobi et al. [15] performed histology on the trend of children's nephrotic syndrome in Ibadan, Nigeria. Since the research into the identification of focal-segmental sclerosis has presented few findings, it is necessary to conduct preprocessing on the image dataset by rotating, changing picture size and replicating images to achieve uniformity. Since the data train is more accurate, this technique is suited for the testing the focal-segmental class. As a result, the information gathered is split into three categories.

Data preparation was performed using previously acquired data, according to the most recent data source. In this step, the transfer learning approach requires data labeling on the dataset to identify the different classes. The HuBMAP dataset was used to label the data train encoded into the path annotation, where the primary data was an image file in the Big TIFF format that was then examined on the raw tiles using Python programming. The Mendeley dataset consisted of 31 SVS pictures that were provided as raw data and converted to PNG files. Data labeling was created for focal-segmental sclerosed, and the data was separated into train, test and validation. Given the variety of techniques used, we considered executing the code offline using the system specified in Table 1 (which was compatible with the Python version) to conduct all tests, as well as performing the experiments using the cloud service using GPU instances.

Table 1. System requirements for running the experiments.

System	Name	Specification	Description
Operating System	Ubuntu	20.04	latest version
Display	NVIDIA	Geforce GTX 1080 Tii	2 GPUs 12 GiB
Memory	DIMM	DDR4 16 GiB	Synchronous 266 Mhz
Harddrive	ATA Disk	1863 GiB (2 TB)	1.7 TB available

Minimum system requirements for conducting the experiments.

The script code was obtained from past research and experiments, and its structure and methods were modified to meet our study goals. The goal of the project was to assess the performance of the experiments and compare them to the dataset that was chosen. As a result, the purpose of this research was to test and enhance the accuracy of picture categorization using the existing image dataset.

2.2. Exploration Data Analysis

The presence or absence of the ring was determined by assessing the circular shape of Bowman's capsule in the nucleus of the glomerular cells. We used the breadth of the red blood cell distribution (RSVP) to distinguish between sclerosed and localized segmental illness, in addition to the study's main emphasis. In the training data portion, this picture recognition method is examined first.

The utilized dataset, as indicated in the preceding section, is an unlabeled dataset collected from three sources. It was then separated into two halves for use as training and test sets. Additionally, one component created a validation set to demonstrate the viability of training or testing according to a predefined categorization. As a result, this study developed a normal, sclerosed and focused training, test and validation set. Figure 2 depicts the validation set data, which includes the true label for each class presented.

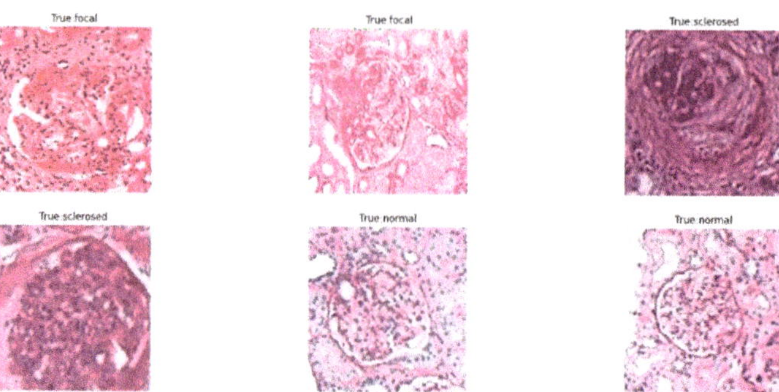

Figure 2. Sample Images of the validation set for the focal-segmental, sclerosed and normal glomeruli.

With a scale of 3:3:1, the data set was split into the train, test and validation sets. The train set had 2100 photos, with 700 pictures for each class, 2290 test sets with 763 pictures for each class and 705 pictures validation set with 235 pictures for each class. Figure 3 depicts the findings, which include a graph of each created set. The training set and validation set as the pre-trained models in the transfer learning process, are generalized based on the dataset's objectives.

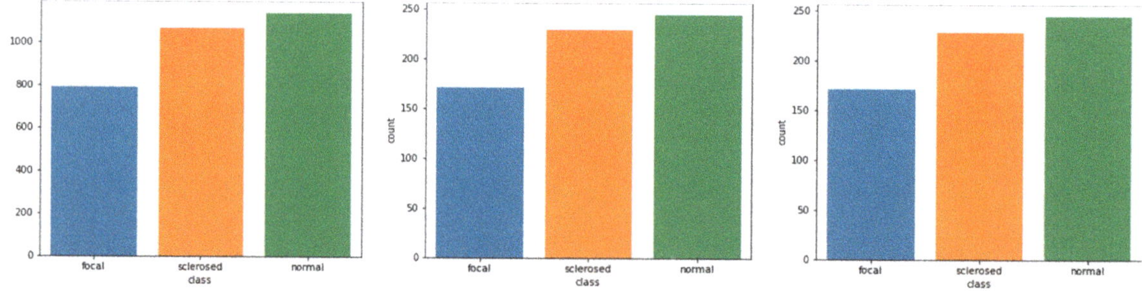

Figure 3. Distribution of experimental data. (**Left**) Distribution of the training set; (**Center**) distribution of the test set and (**Right**) distribution of the validation set.

The dataset utilized must be ignored since the data is generalized; therefore, the image is shrunk into 150 × 150 pixels before the data training process. This step has an influence on the resolution of the loaded image, so make sure it meets the criteria for intensity. The count number ranges from 0 to 8000, and the beginning pixel value is on a scale from 0 to 1. Figure 4 depicts an example of the picture data that meets the learning process's criteria. The example image does not reflect all image intensity values from the dataset, but it does confirm that the data imported is correct.

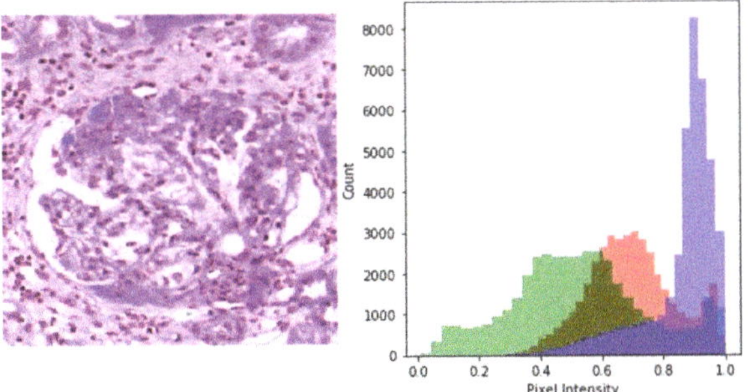

Figure 4. Pixel intensity of images.

2.3. *Efficient Network (EfficientNet)*

Since it has fewer work parameters to optimize the time resources needed, the EfficientNet design necessitates considerable relevance in data processing performance [9]. EfficientNet is comparable to the MobileNetV3 model [16], which uses the MBConv mobile inverted bottleneck as the fundamental building component of its design. The squeeze-excitation layer in the process [17], separates this model from prior versions. Before the picture is extracted further, Figure 5 reveals its resemblance to the previously generated layers and combined additional model to the last layer for obtaining improved results. The schema of the proposed combination model takes the form of a dense layer, in which it is possible to have more of the model than the bundle layer. Afterwards, the design was created after the transfer learning process on each model was completed, since this combination needs the first trained model to process the whole model in one design. However, because this layering issue was still extremely parametric, the impact resulted in a decreased efficiency. Furthermore, as compared to the conventional layers, the combination with depth-wise separated convolution reduces computation by a factor of k2. When the method was presented, the difference was obvious owing to a mix of compound scaling between the scaling width, depth and resolution, with the aim of improving the overall performance using the available resources.

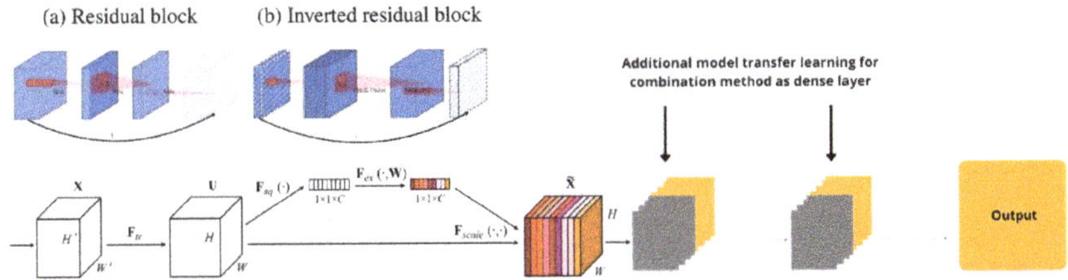

Figure 5. Architecture of the proposed model. (**Top**) Large block placement application in exchange for small blocks in EfficientNet; (**Bottom-Left**) squeeze-excitation is applied to the learning process by adjusting the model used for the input scale data. [6], and (**Bottom-Right**) the dense layer is an additional learning model combined with the main model to produce the learning output.

Compound scaling has two stages: finding the scalping dimension parameters of the baseline network on the resource input using a grid search, and applying the coefficients obtained from adjusting the input dimensions on the baseline network to influence the size of the target model or computational budget, using the coefficients obtained from adjusting the input dimensions on the baseline network. Tan et al. [9] found a mathematical equation that employs the compound coefficient to equalize the scale throughout the network breadth, depth and resolution, as shown below:

$$\text{depth} :\rightarrow d = \alpha^{\phi}$$
$$\text{width} :\rightarrow w = \beta^{\phi}$$
$$\text{resolution} :\rightarrow = \gamma^{\phi}$$
$$\text{s.t.} \rightarrow \alpha{\cdot}\beta^2{\cdot}\gamma^2 \approx 2$$
$$\alpha \geq 1, \beta \geq 1, \gamma \geq 1$$

where the variables α, β, γ are the constants obtained from a small grid search. Logically, ϕ is a user-controlled coefficient regarding the number of resources available in model scaling, where α, β, γ determine the additional resources at the network width, depth and resolution, respectively. In particular, the floating-point operations per second (FLOPS) of the convolution process was proportional to the variables d, w^2, r^2 by doubling the network depth, even though the network width and resolution increased four times. Scaling ConvNet will increase FLOPS by a total of $\left(\alpha \cdot \beta^2 \cdot \gamma^2\right)^{\phi}$ because convolution operations usually dominate the computeing costs presented on ConvNet. EfficientNet contains the same architectural components as the convolutional network in general, despite its distinct working technique.

The EfficientNet-B7 [18] and EfficientNet-L2 [19] models from the EfficientNet architecture are used in this research. These two models are EfficientNet's final models, which have a higher level of precision and scalability than previous EfficientNet models.

2.4. Residual Network (ResNet)

We picked the ResNet model as one of the models in this experiment, as a contrast to the approach we evaluated. ResNet, being the most widely used technique, requires layers to be reformulated as learning residual functions that refer to the layer inputs rather than the learning non-referenced functions [20]. ResNet has an advantage over other models, in that it does not add many layers to directly match the underlying mapping. In one of the examples, ResNet piled the leftover blocks at the top of each network form, illustrating the transfer of the layer work to the next processed layer in ResNet-50, which has 50 layers.

Formally, the function that represents how ResNet works is $\mathcal{F}(x) := \mathcal{H}(x) - x$, where $\mathcal{H}(x)$ is underlying mapping. The original mapping was recast into $\mathcal{F}(x) + x$. There is empirical evidence that these networks are easy to manipulate and optimize, and can gain accuracy by considering the addition of depth to the network without creating new layers.

This research uses the ResNet101V2 and ResNet50V2 models. The equation for this model is to utilize 50 remaining blocks so that the learning process does not take as long to estimate. The only difference between these two models is how the mappings are identified in the learning process [21]. ResNetV2 has this capacity by evaluating the form of the mapping before stacking the rest of the blocks. To put it another way, the model architecture changes as a result of the process. Constant scaling, exclusive gating and shortcuts, such as convolutional or dropout shortcuts, are available in some situations. As a result, while performing the tests, this difference had a substantial impact.

2.5. Very Deep Convolutional Networks for Large-Scale Image Recognition (VGG)

Furthermore, we chose the VGG model as a comparison model in this study, based on the numerous studies that used it in recent years. VGG [22] is a common picture classification method. The VGG's architecture adds depth to the network layer by using a modest network width. The network employs tiny 3×3 filters, where the layer components are made out of the same three blocks, as is standard on CNN. A network, on the other hand, is defined by its simplicity and is otherwise organized with additional components, such as pooling layers and completely linked layers [23]. The VGG16 and VGG19 models were used in this VGG approach.

These two models are the VGG method's final models, with a higher readout rate for the image processing than the other models. VGG16 has 16 weight layers, each of which consists of 13 3 × 3 convolutional layers and 3 fully connected layers, whereas VGG19 has 19 weight layers, each of which consists of 16 3 × 3 convolutional layers and 3 fully connected layers. The VGG19 model differs in that it adds a 3 × 3 convolutional layer to each max pool block with varying sizes, depending on the max pool block.

When compared to numerous newer techniques, the VGG method's performance is deemed steady. This outcome was demonstrated by the method's adaptability, which allowed new methods to be created [8]. The simplicity of the VGG approach influenced other methods that could enhance analytical findings without entirely altering the existing architecture.

3. Results

3.1. Independent Model Experiment Results

We used several treatments for each model in a transfer learning procedure. When presented with the same treatment, the differing model designs caused problems in the learning process. The study began with a comparison of two identical models in the same architecture, with the historical correctness of the models being assessed (Table 2)

Table 2. Performance evaluation of the independent models.

Model Name	Weight	Precision (avg)	Recall (avg)	F1-Score (avg)	Accuracy	Train Estimated Time
VGG19	ImageNet	0.94	0.94	0.93	93.62%	9 m 23 s
ResNet101V2	ImageNet	0.98	0.98	0.98	98.16%	7 m 19 s
EfficientNetB7	ImageNet	0.95	0.95	0.95	94.61%	12 m 52 s
EfficientNetL2	Noisy student	0.11	0.33	0.17	33.33%	12 m 36 s

VGG19. VGG19 was tested by training a 3 × 3 convolutional layer using an ImageNet classification model using average pooling. The output was assessed by permitting numerous fine-tuning approaches, such as the activation of the rectified linear unit (ReLU) dense layer to achieve a low learning rate, the activation of the soft-max layer and saving checkpoints on the best model [24]. The period began with an evaluation of the time spent during training, the number of total parameters processed and the storage of a learning process evaluation. With a total parameter of 20,090,435, VGG19 successfully completed the training with an average model accuracy of 0.9361, validated with a high value of 0.9390 and the maximum accuracy value of 0.9981. It is possible to infer that the two VGG models can depict models that are good enough to be classed as medical pictures, based on the findings of the two models. The accuracy of the case suggested by the VGG model was in the range of 0.80 to 0.90 and above. In terms of the value scale, the VGG19 value is an appropriate category to use as an image classification model, as shown in Table 3. Figure 6a demonstrates that the original value's correlation with the predicted value shows a high color prediction for the focal and sclerosed classes, while the projected normal class has a sufficient color in the classification.

Table 3. Performance evaluation combined models.

Model Name	Precision (avg)	Recall (avg)	F1-Score (avg)	Accuracy	Estimated Time
VGG16 + VGG19	0.94	0.94	0.93	93.62%	21 m 25 s
ResNet101 + VGG16 + VGG19 + EfficientNetL2	0.95	0.95	0.95	94.89%	40 m 38 s
ResNet101V2 + EfficientNet-L2	0.97	0.97	0.97	97.16%	19 m 55 s
ResNet101V2 + VGG16 + EfficientNetB7	0.97	0.97	0.97	97.02%	33 m 13 s

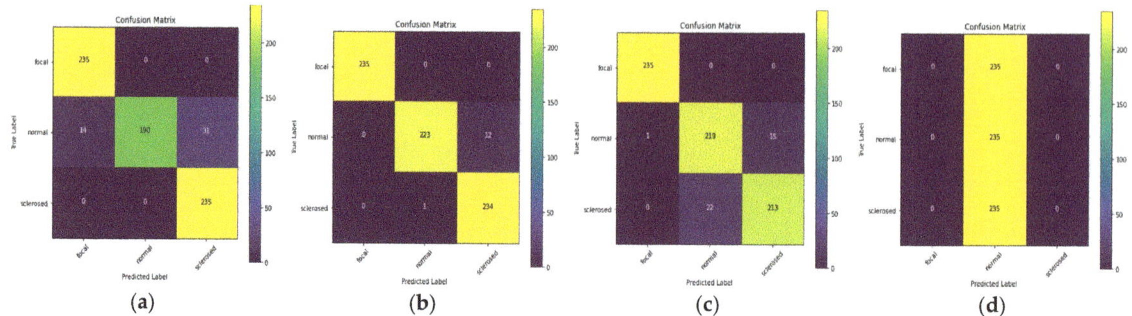

Figure 6. Confusion matrixes of various models. (**a**) VGG19; (**b**) ResNet101V2; (**c**) EfficientNet-B7 and (**d**) EfficientNet-L2.

ResNet101V2. The architectural resemblance between these two models may also result in the same model performance configuration. On the convolutional network, both model sets are constructed with a resolution of 150 × 150 using ImageNet weight. With a logistic regression (LR) value of 0.01 and a momentum value of 0.7, both of these models employed stochastic gradient descent (SGD) optimization for fine-tuning [25]. This setting is an iterative approach for fine-tuning the objective function. Each high-accuracy validation point was saved in the model checkpoint function and utilized as a weight model in a pre-trained model.

ResNet101V2 runs 55,637,123 parameters in total. After determining these parameters, ResNet101V2 completed the training with an average accuracy of 0.9815, which was confirmed by the maximum value of 0.9801, for which the greatest accuracy score was 1000. The training results, as shown in Table 3, demonstrate that each model employed in the experiment had the best accuracy.

EfficientNetB7 and EfficientNetL2. The EfficientNet model employed in this study differs significantly from the previous one. EfficientNet-B7 can operate using ImageNet weight, but EfficientNet-L2 requires noisy student weight as a pre-trained model, according to prior research. This occurred because EfficientNet-L2 was incompatible with the ImageNet weights, when the number of layer weight initiations of the model differed. The unevenness of the pre-trained models employed was caused by weight fluctuations, although this is an exception to correctly performing the training process. Even when running on different models, these two models employed the same fine-tuning. The average-pool layer was added to the end of the flattened layer. In addition, the use of the ReLU feature and a dropout layer with a value of 0.2 created a dense layer. A thick layer was seen in the last portion. Furthermore, this approach employed an optimizer in the form of Adam [26], throughout the compilation process.

With a training parameter of 2,625,539, EfficientNet-B7 achieved an average accuracy of 0.9461. The highest value, 0.9291, was used to confirm this accuracy, and the maximum accuracy was 0.9662. Meanwhile, EfficientNet-L2 utilized a training parameter of 5,640,195, with an average accuracy of 0.3333 as a consequence. The greatest accuracy as 0.3424, and this accuracy was confirmed using the highest value of 0.3333.

The EfficientNet-B7 and EfficientNet-L2 designs had significant variations in their training outcomes. EfficientNet-B7 does a far better job at presenting results than EfficientNet-L2. Figure 6c,d indicate that the EfficientNet-L2 predictions show that everything is in the usual class, but the EfficientNet-B7 predictions are evenly distributed in each class. These findings demonstrate that the allied model does not generate accurate transfer learning predictions.

Based on the training data, we may infer that the EfficientNet model does not have a high level of accuracy. The inequality found in the EfficientNet-L2 model indicates that the accuracy findings in EfficientNetB7 are inconsistent. The model's incompatibility with the dataset utilized, on the other hand, prevents it from producing the optimal results.

3.2. Combining Model Experiment Results

After reviewing the results of the experiment, we tried again, this time using many models that were judged unfit for classification. As shown in Table 3, the findings of Effi-cientNet-L2 do not exhibit a high degree of accuracy, especially when compared to the criteria set by the trials. In a model with low accuracy, a high yield model has a substantial impact on improving accuracy. The experiments on a cognate or unrelated model confirmed this idea.

Combination Allied Model. The accuracy value obtained from the combination of the associated models is dependent on the base model utilized. The value obtained in the combination model was proportional to the model's independent accuracy value. This experiment was carried out on three family models that is VGG, ResNet, and EfficientNet, all of which have different layer designs despite being related. With VGG16 as the basic model, the VGG model accuracy is 93.62 percent, as shown in Table 4. According to Figure 7, the normal class distribution offers less predictive data than the focused and sclerosed classes.

Table 4. Classification of the allied model.

Classify	Evaluation		
	Precision	Recall	F1-Score
Focal	0.94	1.00	0.97
Normal	1.00	0.81	0.89
Sclerosed	0.88	1.00	0.94

VGG16 and VGG19 combined into the combination model produce better results.

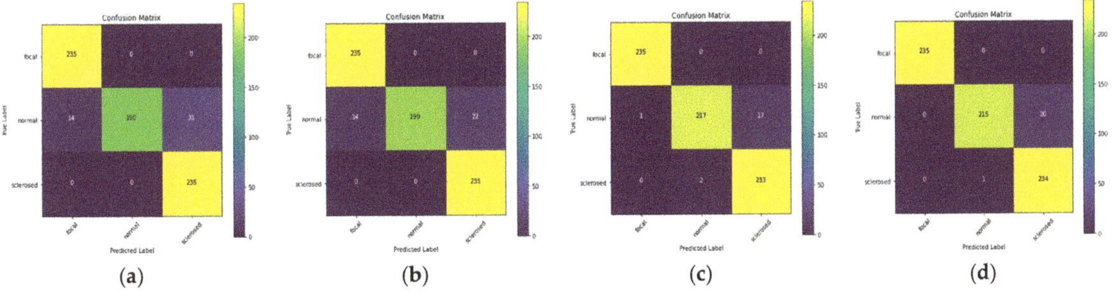

Figure 7. Confusion matrixes of combined models. (**a**) Combined model of VGG16 and VGG19 (allied model); (**b**) combined model of ResNet101, VGG16, VGG19 and EfficientNetL2 (multivariate model); (**c**) combined model of ResNet101V2 and EfficientNet-L2 (cross model) and (**d**) combined model of ResNet101V2, VGG16 and EfficientNetB7 (multivariate model).

Combination Cross-Model. Cross-model combination refers to the model that is prioritized in order to enhance its accuracy value, limiting the models that may be combined to those that meet these requirements. EfficientNet-L2 and ResNet 101, for example, confirmed the need to improve. As a result, it is necessary to properly test each of these two models.

According to the base model employed in these tests, there are some varied categorization findings. Table 4 indicates that the classification results of Res-Net101V2 may enhance the accuracy results of EfficientNet-L2 by approaching the ResNet101V2 value independently, by achieving 97.16 percent. The projected value in Figure 7c is affected by this finding, with the distributions of each class matching the original value with a few erroneous values.

Combination Multivariate Model. EfficientNet-L2 has risen by up to two times while using ResNet101 as the basic model, compared to the prior trial when the model was conducting independent training. In this situation, ResNet101 has a low accuracy value, but it is still better than EfficientNet-L2; thus, the gain is not substantial. As a result of attempting to add another model with greater accuracy to this combination, two VGG models are used in the classification process as a multi-model combination experiment. The accuracy value from the combination model produces similar results, which are better than the EfficientNet-B7 model independently.

VGG, ResNet and EfficientNet were all used to perform model tests. One model from each model family was selected as the basic model, since it generated high accuracy results. Each model's accuracy has greatly improved as a result of this setup. This model's combination experiment takes two to three times as long as the models that conduct experiments individually, depending on the number of models that can be combined. As shown in Table 4, the time necessary to combine three particular models to obtain the best results, is equal to the training time for each model. This impact does not apply to the combinations inside the same model, since the predicted outcomes do not need numerous iterations, allowing the results to be taken from the same model without having to retrain the model. However, by combining the two models in the same model, it does not significantly enhance the model classification accuracy. As a result, in order to obtain more optimum findings, the time efficiency of this model combination experiment must be addressed.

3.3. *Evaluation*

We compared and contrasted two categorization studies, each of which had its own set of benefits and drawbacks. This study discovered mixed findings from all of the models evaluated in the different model trials. These outcomes were influenced by the model architecture used, as well as the fine-tuning the model. The projected time for each model was also dependent on the parameters that were utilized during training; the more parameters that are used, the longer the training will take. Furthermore, the training per epoch method produced diverse graphs. The graphs observed did not develope steadily and there was significant irregularity in the graphs acquired. As a result, it is not recommended to use a model with such dramatic findings for classifying medical pictures. The F1-score evaluation parameter was utilized to examine the value of each class and the mean of each model. The accuracy and recall levels in the arithmetic theorem provided this F1-score. Table 5 indicates that the normal glomeruli class dominates the EfficientNet-L2 model's prediction results, which is consistent with Figure 6d, which displays predictions in the normal class. In Table 6, the class predictions for ResNet101V2 are dispersed in each class with a modest error rate, as shown in Figure 6b, in which the original and predicted values overlap fairly well.

When comparing the independent assessment models displayed in Figure 8 (left), it is clear that EfficientNet-L2 is not acceptable as a reference model for medical image classification since it has the lowest evaluation value, with just a 0.17 F1-score. With an F1-score of 0.98, ResNet101V2 is the best reference model for medical picture categorization. The findings of the combination technique experiment were identical for each model combination. Table 4 demonstrates that the model combination experiment yields findings that are more than 90% accurate over time. Figure 8 shows this outcome, with a predictive value in each class's distribution that matches the original value and many error values in the incorrect predictions.

Table 5. Classification of EfficientNetL2.

Classify	Evaluation		
	Precision	**Recall**	**F1-Score**
Focal	0.00	0.00	0.00
Normal	0.33	1.00	0.50
Sclerosed	0.00	0.00	0.00

The classification performance obtained via EfficientNetL2.

Table 6. Classification of Resnet101V2.

Classify	Evaluation		
	Precision	**Recall**	**F1-Score**
Focal	1.00	1.00	1.00
Normal	1.00	0.95	0.97
Sclerosed	0.95	1.00	0.97

The classification performance obtained via Resnet101V2.

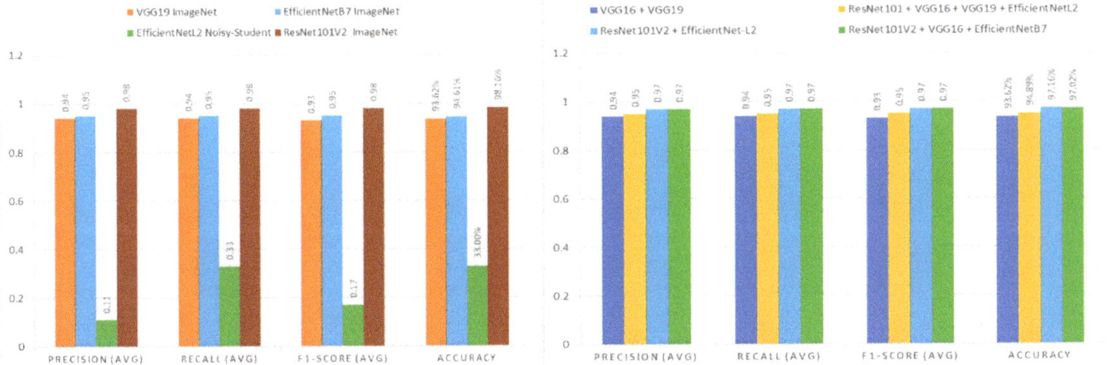

Figure 8. Comparison performance evaluation metrics. (**Left**) Comparison performance independent model and (**Right**) comparison performance combination model. The average of each of the evaluation metrics is obtained from the total evaluation metrics value per epoch divided by the total epoch from each model $\left(\overline{EM} = \frac{\sum EM}{\sum Epoch}\right)$.

Based on these findings, this study assesses each model's classification prediction class to demonstrate the impact of accuracy on model performance. The three model combination trials revealed a metric assessment value that was relatively steady with a low error rate (Tables 7–9). The prediction value of the class distribution, which is near to 1.00, indicates that the forecast value closely follows the actual value. Figure 8 (right) indicates that all the combination technique trials achieved an assessment value of more than 0.90, with an average F1-score of 0.955. However, the predicted training time for this model combination experiment was very high, with an average of 28 min required, making it less time efficient.

Table 7. Classification of the cross model.

Classify	Evaluation		
	Precision	**Recall**	**F1-Score**
Focal	1.00	1.00	1.00
Normal	0.99	0.92	0.96
Sclerosed	0.93	0.99	0.96

The classification performance of the combined model with ResNet101V2 and EfficientNet-L2.

Table 8. Classification of the multivariate model.

Classify	Evaluation		
	Precision	Recall	F1-Score
Focal	1.00	1.00	1.00
Normal	1.00	0.91	0.95
Sclerosed	0.92	1.00	0.96

The classification performance of the combined model with ResNet101V2, VGG16, and EfficientNetB7.

Table 9. Previous works and the performance metrics regarding glomeruli research.

Author Glomeruli Classification	Method	Data	Classes	Name of Classes	Mean Performance Metrics
Bueno et al. [8]	SegNet + VGG19	1245 images	3	Non-Glomeruli, Normal and Sclerosed	81.91% F1-score
Bueno et al. [12]	AlexNet	1245 images	2	Normal and Sclerosed Glomeruli	99.57% F1-score
Altini et al. [7]	DeepLab v3+	2344 images	2	Non-Sclerosed and Sclerosed Glomeruli	84.64% F1-score
Barros et al. [27]	KNN	811 images	2	Normal and Proliferative Glomeruli	88.3% accuracy
Marsh et al. [28]	Fully CNN	3867 images	3	Tubulointerstitial, Non-Sclerosed and Sclerosed	84.75% F1-score
Kannan et al. [14]	InceptionV3	1496 images	3	No glomerulus, Normal or Partially Sclerosed, and Globally Sclerosed	92.67% accuracy
Proposed Research Method	Combined Model	3924 images	3	Normal, Sclerosed and Focal Glomeruli	98.16% F1-score

Similar research related to chronic kidney disease base on glomeruli disease.

4. Discussion

Based on the research results, the combination of deep transfer learning methods to classify kidney disease in glomeruli has many characteristics, in terms of technical data collection, exploratory data analysis and also the comparison with relevant research. By prioritizing the data obtained from previous research coupled with the latest data from relevant health institutions, we attempted to filter the data that was suitable and with a high degree of similarity, so that the data collected could improve the results of machine learning. This technique was also carried out in previous research, by Manzo et al. [29], which used CX-ray images data from various sources to detect COVID-19 disease through machine learning. However, the approach of this study used segmentation techniques from diseases images; therefore, its steps are quite tedious. Meanwhile, we used the diseased kidney data as a dataset to be studied by the machine. As a result, when testing the data, the machine already recognized the exact criteria for the diseased kidney.

For our data exploration, we focused on determining the level of the distribution of red blood cells (RSVP) that appeared in the diseased kidney image by adjusting the pixels and brightness of the image, so that it appeared more proportional. This technique is in line with the method used by Pavinkurve et al. [30], which performs image preprocessing to detect the same disease. In this study, this technique was able to overcome the problem of image bias in CT scan images. The suitability of this technique with the results of the CT scan images is considered as being capable of facilitating the machine learning process to obtain the object of the image. On the other hand, this study utilizes subject matter that is often studied in previous studies, but is unique in its application of the methods.

This research, which focuses on discussing kidney disease, has relevant developments with the same objectives, including the utilization of image classification that applies different machine learning methods in various amounts, and the segmentation of images by combining classification and segmentation methods. We obtained this perspective by developing a combined model research method that utilized various types of methods to compare which results were more accurate and precise, as was also performed by Bueno et al. in two kidney disease studies, which combined the resulting segmentation method with the adopted classification method to find the significant differences in diseased kidneys. Furthermore, they found a new method to produce a more accurate optimal image classification method for diseased kidneys. By utilizing the results obtained in the research of Bruno et al., we expanded the research by adding various transfer learning methods that had been previously studied and adding to the dataset used, so that the results obtained were compared with each other to achieve a higher accuracy value.

The technique that we found in this study is also supported by previous research, which uses a variety of techniques. The use of automatic learning in kidney disease is our focus for the development of the current research. This is in line with what was conducted by Altini et al. [7], who developed research on kidney disease using the DeepLab method through MATLAB, whereby the learning process can present the detailed results requested by the user to facilitate comprehensive research performance. Meanwhile, several previous studies used neural network techniques, either by directly using the model or installing a new model. This technique was found in a study developed by Barros et al. [27] and Marsh et al. [28], by installing neural networks to find the types of glomerular proliferative kidney disease, which implements neural networks in the form of a full model to detect the type of tubulointerstitial kidney glomeruli disease. This condition prompted us to conduct research on a combination of models that utilize neural networks in the transfer learning model. Thus, we could develop our research based on the learning scheme of neural networks in each of the models we used. In addition, the research we used was also supported by the research of Kannan et al. [14], who researched the differences in the sclerosed shape of diseased kidney glomeruli using the CNN model in the form of Inception V3.

5. Conclusions

To obtain the desired results, medical image categorization necessitates the following procedures. The fundamental task in deep machine learning is to find an appropriate model to run on a medical imaging dataset, which may be accomplished by altering an existing model, developing a new model or applying an existing model. This study compared each model to obtain a high classification accuracy value by using an existing model on a medical picture dataset. This study integrated many models into a single model to obtain a higher and more consistent classification accuracy value from the current models.

We found the EfficientNet model to be weak in comparison to various other machine learning models after analyzing the suggested technique. This study demonstrated that EfficientNet-L2 is not yet ideal for correctly identifying medical pictures, since it only achieves a 33 percent accuracy rate. As a result, the EfficientNet model is ineffective in classifying medical images independently. However, we discovered that the ResNet model is a good fit for categorizing medical images. We discovered a ResNet101V2 that is appropriate for usage in the medical picture dataset gathered during the application of current models. ResNet101V2 has the best accuracy of all the tests performed on an independent model, with values over 98.16 percent. As a result of this finding, it is clear that certain models are less accurate than others, necessitating the use of alternative techniques. As a result, combining models becomes a new experiment in order to improve the accuracy of the unfavorable model.

The allied, cross and multivariate models were evaluated in this combination experiment. The classification results from the three trials demonstrate that by using the best model as the base model, all combination models attain accuracy values above 90%. This application focuses on the best model in order to enhance the less-than-ideal model. Since it works highly depending on the number of merged models, the combined model has flaws in terms of the computation time. However, when compared to the models created separately, the absolute accuracy of the combined model is the highest.

With the advancement of medical image classification research, it is important to address the various aspects that influence the model employed. Existing models may be improved by inputting or altering various optimization variables to obtain the best results for the combined model, as well as the procedures required to streamline the model process's estimated time and its impact on the model architecture. To put it another way, this research is anticipated to continue discovering the best outcomes in terms of architecture and the predicted time necessary, so that the accuracy value attained is achieved using the best methods and models.

Author Contributions: Conceptualization, H.-C.L. and A.F.A.; methodology, H.-C.L.; software, A.F.A.; validation, H.-C.L. and A.F.A.; formal analysis, H.-C.L.; investigation, A.F.A.; resources, A.F.A.; data curation, A.F.A.; writing—original draft preparation, A.F.A.; writing—review and editing, H.-C.L.; visualization, A.F.A.; supervision, H.-C.L.; project administration, H.-C.L.; funding acquisition, H.-C.L. All authors have read and agreed to the published version of the manuscript.

Funding: This research was funded partially by the Ministry of Education of Taiwan (R.O.C.) via the Artificial Intelligence Talent Cultivation Project and was supported partially by the Ministry of Science and Technology of Taiwan (R.O.C) through National Center for High-performance Computing.

Institutional Review Board Statement: Not applicable.

Informed Consent Statement: Not applicable.

Data Availability Statement: Not applicable.

Acknowledgments: We thank Khrisna Kumar S for their input in the project. This program code is modified from Kaggle to suit the needs of the dataset and the architecture used. This research is based on an image database supported by the Cancer Imaging Archive. This work makes use of the supercomputing facilities managed by the Artificial Intelligence Laboratory, National Quemoy University and the online facilities from Ministry Science and Technology of Taiwan through Taiwan Cloud Computing Preferences. This research thanks Google for providing a valuable source of data for our research.

Conflicts of Interest: The authors declare no conflict of interest.

References

1. Duchnowski, P.; Hryniewiecki, T.; Kuśmierczyk, M.; Szymański, P. The usefulness of selected biomarkers in patients with valve disease. *Biomarkers Med.* **2018**, *12*, 1341–1346. [CrossRef] [PubMed]
2. Duchnowski, P.; Hryniewiecki, T.; Kuśmierczyk, M.; Szymański, P. Anisocytosis predicts postoperative renal replacement therapy in patients undergoing heart valve surgery. *Cardiol. J.* **2020**, *27*, 362–367. [CrossRef] [PubMed]
3. George, S.; Pal, A.C.; Gagnon, J.; Timalsina, S.; Singh, P.; Vydyam, P.; Munshi, M.; Chiu, J.E.; Renard, I.; Harden, C.A.; et al. Evidence for SARS-CoV-2 spike protein in the urine of COVID-19 patients. *Kidney* **2021**, *2*, 924–936. [CrossRef]
4. Kaur, S.P.; Gupta, V. COVID-19 Vaccine: A comprehensive status report. *Virus Res.* **2020**, *288*, 198114. [CrossRef] [PubMed]
5. Kolhe, N.V.; Fluck, R.J.; Selby, N.M.; Taal, M.W. Acute kidney injury associated with COVID-19: A retrospective cohort study. *PLoS Med.* **2020**, *17*, e1003406. [CrossRef] [PubMed]
6. Rosenberg, A.; Kopp, J. Focal segmental glomerulosclerosis. *Clin. J. Am. Soc. Nephrol.* **2017**, *12*, 502–517. [CrossRef] [PubMed]
7. Altini, N.; Cascarano, G.D.; Brunetti, A.; Marino, F.; Rocchetti, M.T.; Matino, S.; Venere, U.; Rossini, M.; Pesce, F.; Gesualdo, L.; et al. Semantic segmentation framework for glomeruli detection and classification in kidney histological sections. *Electronics* **2020**, *9*, 503. [CrossRef]
8. Bueno, G.; Fernandez-Carrobles, M.M.; Gonzalez-Lopez, L.; Deniz, O. Glomerulosclerosis identification in whole slide images using semantic segmentation. *Comput. Methods Programs Biomed.* **2020**, *184*, 105273. [CrossRef] [PubMed]
9. Tan, M.; Le, Q.V. Efficientnet: Rethinking model scaling for convolutional neural networks. In Proceedings of the 36th International Conference on Machine Learning ICML, Jacksonville Beach, FL, USA, 11 June 2019; pp. 10691–10700.
10. Foret, P.; Kleiner, A.; Mobahi, H.; Neyshabur, B. Sharpness-aware minimization for efficiently improving generalization. *arXiv* **2020**, arXiv:2010.01412.
11. Martín-Navarro, J.A.; Gutiérrez-Sánchez, M.J.; Petkov-Stoyanov, V.; Justo-Ávila, P.; Ionela-Stanescu, R. Glomerulonefritis focal y segmentaria en paciente con sarcoidosis pulmonar. *Nefrol* **2013**, *33*, 431–433.
12. Bueno, G.; Gonzalez-Lopez, L.; Garcia-Rojo, M.; Laurinavicius, A.; Deniz, O. Data for glomeruli characterization in histopathological images. *Data Brief* **2020**, *29*, 105314. [CrossRef] [PubMed]
13. Dimitris. HuBMAP: Glomeruli Tiles and Mosaics, Kaggle. 2020. Available online: https://www.kaggle.com/anadelta/hubmap-glomeruli-tiles (accessed on 17 March 2021).
14. Kannan, S.; Morgan, L.A.; Liang, B.; Cheung, M.G.; Lin, C.Q.; Mun, D.; Nader, R.G.; Belghasem, M.E.; Henderson, J.M.; Francis, J.M.; et al. Segmentation of glomeruli within trichrome images using deep learning. *Kidney Int. Rep.* **2019**, *4*, 955–962. [CrossRef] [PubMed]
15. Asinobi, A.O.; Ademola, A.D.; Okolo, C.A.; Yaria, J.O. Trends in the histopathology of childhood nephrotic syndrome in Ibadan Nigeria: Preponderance of idiopathic focal segmental glomerulosclerosis. *BMC Nephrol.* **2015**, *16*, 213. [CrossRef] [PubMed]

16. Howard, A.; Sandler, M.; Chen, B.; Wang, W.; Chen, L.-C.; Tan, M.; Chu, G.; Vasudevan, V.; Zhu, Y.; Pang, R.; et al. Searching for MobileNetV3. In Proceedings of the International Conference on Computer Vision, Seoul, Korea, 27 October–2 November 2019; pp. 1314–1324. [CrossRef]
17. Hu, J.; Shen, L.; Albanie, S.; Sun, G.; Wu, E. Squeeze-and-excitation networks. *IEEE Trans. Pattern Anal. Mach. Intell.* **2020**, *42*, 2011–2023. [CrossRef] [PubMed]
18. Xie, Q.; Luong, M.-T.; Hovy, E.; Le, Q.V. Self-training with noisy student improves imagenet classification. In Proceedings of the 2020 IEEE/CVF Conference on Computer Vision and Pattern Recognition (CVPR), Seattle, WA, USA, 13–19 June 2020; pp. 10684–10695.
19. Pham, H.; Dai, Z.; Xie, Q.; Le, Q.V. Meta pseudo labels. *arXiv* **2020**, arXiv:2003.10580.
20. He, K.; Zhang, X.; Ren, S.; Sun, J. Deep residual learning for image recognition. In Proceedings of the 2016 IEEE Conference on Computer Vision and Pattern Recognition (CVPR), Las Vegas, NV, USA, 27–30 June 2016; pp. 770–778. [CrossRef]
21. He, K.; Zhang, X.; Ren, S.; Sun, J. Identity mappings in deep residual networks. In *Computer Vision—European Conference on Computer Vision*; Springer: Cham, Switzerland, 2016; pp. 630–645. Available online: https://link.springer.com/chapter/10.1007/978-3-319-46493-0_38 (accessed on 23 February 2021).
22. Zhu, J.; Shen, B.; Abbasi, A.; Hoshmand-Kochi, M.; Li, H.; Duong, T.Q. Deep transfer learning artificial intelligence accurately stages COVID-19 lung disease severity on portable chest radiographs. *PLoS ONE* **2020**, *15*, e0236621. [CrossRef] [PubMed]
23. Simonyan, K.; Zisserman, A. Very deep convolutional networks for large-scale image recognition. In Proceedings of the 3rd International Conference on Learning Representations, ICLR 2015, San Diego, CA, USA, 7–9 May 2015; pp. 1–14.
24. Javid, A.M.; Das, S.; Skoglund, M.; Chatterjee, S. A ReLU dense layer to improve the performance of neural networks. In Proceedings of the ICASSP 2021–2021 IEEE International Conference on Acoustics, Speech and Signal Processing (ICASSP), Toronto, ON, Canada, 6–11 June 2021; pp. 2810–2814. Available online: http://arxiv.org/abs/2010.13572 (accessed on 3 March 2021).
25. Sra, S.; Nowozin, S.; Wright, S.J. *Optimization for Machine Learning*; MIT Press: Cambridge, MA, USA, 2012.
26. Kingma, D.P.; Ba, J. Adam: A method for stochastic optimization. In Proceedings of the 3rd International Conference on Learning Representations, ICLR 2015, San Diego, CA, USA, 7–9 May 2015; Available online: https://arxiv.org/abs/1412.6980#:~{}:text=WeintroduceAdam%2C (accessed on 4 February 2021).
27. Barros, G.O.; Navarro, B.; Duarte, A.; Dos-Santos, W.L.C. PathoSpotter-K: A computational tool for the automatic identification of glomerular lesions in histological images of kidneys. *Sci. Rep.* **2017**, *7*, srep46769. [CrossRef] [PubMed]
28. Marsh, J.N.; Matlock, M.K.; Kudose, S.; Liu, T.-C.; Stappenbeck, T.S.; Gaut, J.P.; Swamidass, S.J. Deep learning global glomerulosclerosis in transplant kidney frozen sections. *IEEE Trans. Med Imaging* **2018**, *37*, 2718–2728. [CrossRef] [PubMed]
29. Manzo, M.; Pellino, S. Fighting together against the pandemic: Learning multiple models on tomography images for COVID-19 diagnosis. *AI* **2021**, *2*, 261–273. [CrossRef]
30. Pavinkurve, N.P.; Natarajan, K.; Perotte, A.J. Deep vision: Learning to identify renal disease with neural networks. *Kidney Int. Rep.* **2019**, *4*, 914–916. [CrossRef] [PubMed]

Article

Internet of Things-Driven Data Mining for Smart Crop Production Prediction in the Peasant Farming Domain

Luis Omar Colombo-Mendoza [1], Mario Andrés Paredes-Valverde [1], María del Pilar Salas-Zárate [1,*] and Rafael Valencia-García [2]

[1] Tecnológico Nacional de México/I. T. S., Teziutlán 73960, Mexico; luis.cm@teziutlan.tecnm.mx (L.O.C.-M.); mario.pv@teziutlan.tecnm.mx (M.A.P.-V.)
[2] Departamento de Informática y Sistemas, Campus de Espinardo, Universidad de Murcia, 30100 Murcia, Spain; valencia@um.es
* Correspondence: maria.sz@teziutlan.tecnm.mx

Abstract: Internet of Things (IoT) technologies can greatly benefit from machine-learning techniques and artificial neural networks for data mining and vice versa. In the agricultural field, this convergence could result in the development of smart farming systems suitable for use as decision support systems by peasant farmers. This work presents the design of a smart farming system for crop production, which is based on low-cost IoT sensors and popular data storage services and data analytics services on the cloud. Moreover, a new data-mining method exploiting climate data along with crop-production data is proposed for the prediction of production volume from heterogeneous data sources. This method was initially validated using traditional machine-learning techniques and open historical data of the northeast region of the state of Puebla, Mexico, which were collected from data sources from the National Water Commission and the Agri-food Information Service of the Mexican Government.

Keywords: data mining; predictive analytics; Internet of Things; peasant farming; smart farming system; crop production prediction

Citation: Colombo-Mendoza, L.O.; Paredes-Valverde, M.A.; Salas-Zárate, M.d.P.; Valencia-García, R. Internet of Things-Driven Data Mining for Smart Crop Production Prediction in the Peasant Farming Domain. *Appl. Sci.* **2022**, *12*, 1940. https://doi.org/10.3390/app12041940

Academic Editor: Manuel Armada

Received: 27 December 2021
Accepted: 10 February 2022
Published: 12 February 2022

Publisher's Note: MDPI stays neutral with regard to jurisdictional claims in published maps and institutional affiliations.

Copyright: © 2022 by the authors. Licensee MDPI, Basel, Switzerland. This article is an open access article distributed under the terms and conditions of the Creative Commons Attribution (CC BY) license (https://creativecommons.org/licenses/by/4.0/).

1. Introduction

The set of techniques that allow for manually and automatically extracting information that resides implicitly in data in a nontrivial way and that could be useful for various processes is known as data mining [1]. Data mining is rooted in artificial intelligence, particularly, in machine learning (ML), as well as in statistical analysis. Through models extracted using artificial intelligence and statistical analysis techniques it is possible to solve problems that imply prediction, classification and segmentation tasks, meaning that large amounts of data can be processed and used more efficiently [2].

Furthermore, the process of extracting information from large datasets with the aim of making estimations about future results is known as predictive analytics. It represents an intermediate step within a broader process of data analytics known as business analytics [3]. In this context, machine learning can be defined as a data analysis method that automates the construction of analytical models. Its study is based on the idea that software systems can learn at least semiautonomously from information by identifying patterns and making decisions with minimal human intervention.

Predictive analytics, along with Internet of Things (IoT) technologies, has been extensively applied to the agricultural domain in recent years [4–9]. This has enabled the development of the concepts of smart agriculture/farming and precision agriculture/farming. IoT technologies are the set of predominant and emerging Information and Communication Technologies (ICT) that are the foundation of a global infrastructure for the information society, which enables advanced services by interconnecting virtual and physical "things".

According to the Food and Agriculture Organization of the United Nations, agriculture in Mexico represents more than an important productive sector. Beyond its contribution to the national GDP, which is barely 4%, the multiple functions of agriculture in Mexico's economic, social and environmental development indicate that its incidence is much greater than that indicator would imply.

In particular, the volume of agricultural production in the Mexican state of Puebla contributed 7,403,938 tons to the country's agricultural production in 2018, ranking 15th among the 32 states of the country. Nevertheless, this contribution implies 22.6% of the economically active population of the state of Puebla, which shows a disparity that indicates that the Puebla's farmlands are not very efficient.

Considering that the primary sector of the economy is the primary source of food and sustenance for families that live in rural communities, and even in rural communities that are very far from the urban centers in Puebla, it is evident that rural development is one of the main pillars of the growth and well-being of society in Puebla.

This work seeks to contribute to the recovery of the farmland of the Mexican state of Puebla, which is one of the major purposes of its government, by proposing: (1) the architectural design of a smart peasant farming system for crop production prediction, which is based on low-cost IoT sensors and popular data storage services and data analytics services on the cloud, and (2) a new data-mining method exploiting climate and crop production data sources for the prediction of the volume of production of corn grain in the northeast region of the state of Puebla.

Figure 1 provides an overview of our research idea; it formally shows a simplified version of the workflow of the proposed system architecture, in which only the major components, and the interactions among them, are included. This figure highlights the generation of crop production predictions as output, as well as the roles of IoT-based sensors and the peasant farmer (as end user) in provisioning heterogeneous input data.

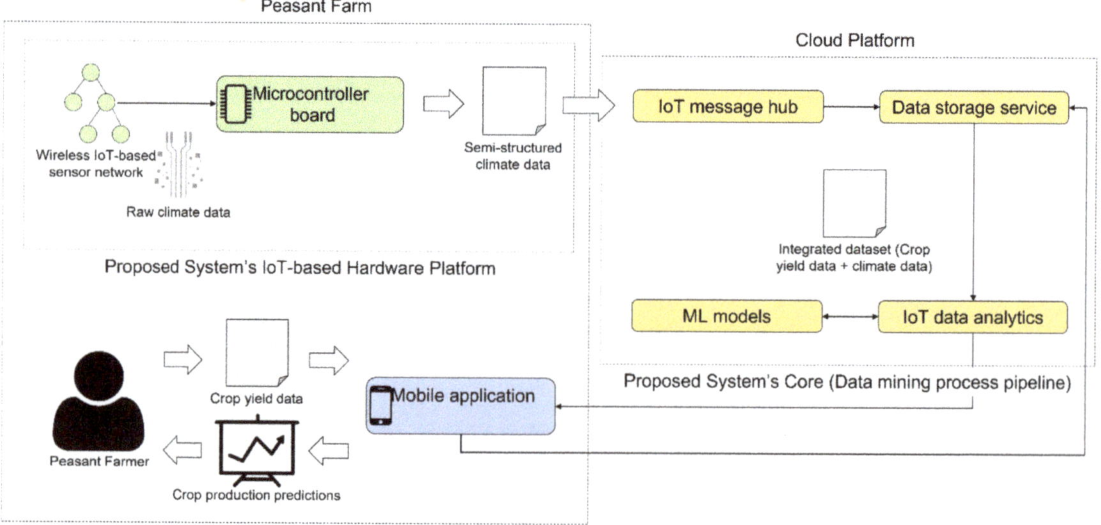

Figure 1. Research Idea Overview.

As shown in Figure 1, climate data, namely, temperature and rainfall, are automatically collected from IoT-based sensors (other climate data, namely, storm activity and fog and hail occurrence, are gathered from datasets made available by local weather stations), whereas crop yield data, namely, hectares planted, hectares harvested and actual crop production volumes, are provided by the peasant farmer though a mobile application. Furthermore, notice that the interactions among the components that represent the core of the system architecture: IoT message hub, data storage service, IoT data analytics service and ML models depict the main steps of the proposed data-mining method. Unlike most recent related works, our work proposes a data-mining process for the prediction of the volume of crop production that integrates two heterogeneous sources of crop production data and climate data as well as the architectural design of a smart farming system specially designed for peasant farming.

The remainder of this paper is organized as follows: in the first section, some recent proposals that are relevant to our work are described and compared; in the second section, our proposal is presented in detail; in the third section, the results of an initial validation of our proposal are presented and discussed; finally, in the last section, the concluding remarks of our work are summarized and the future lines of research are outlined.

2. State of the Art

Some of the state-of-the-art proposals that are more related to our proposal are described below for comparison purposes.

A smart farming system using Internet of Things (IoT) technologies and machine-learning techniques for data mining is presented in [10]. It is based on an architecture comprising the following four layers: (a) a layer consisting of IoT sensors and actuators deployed in the farming field, (b) an edge-computing layer that provides integration between an IoT wireless sensor network and the cloud using IoT gateways, (c) a cloud-computing layer that is intended to store and analyze data over cloud servers and (d) an end user mobile- and web-based application layer. In that work, a new prediction method is proposed as the foundation of a decision support system for crop productivity and drought prediction based on the integration of the PART classification technique and the wrapper feature selection approach.

F. Balducci et al. [11] present five cheap, practical and easy-to-implement data analysis experiments intended to increase smart farming productivity. These experiments range from forecasting of future crop harvest on complete time-series data to reconstruction of missing or wrong IoT sensors data, passing through the detection of faulty IoT sensors from the geographical clustering of source monitoring stations using the Euclidean distance metric. A variety of machine-learning algorithms such as decision tree, k-nearest neighbors and linear regression, as well as a single-layer perceptron neural network, were used and compared for these purposes in conjunction with three heterogeneous datasets belonging to industry, scientific research and statistical institutions.

A precision agriculture system that seeks to reduce efforts and labor of agricultural sector personnel, which uses IoT sensors for data collection, as well as machine-learning and deep-learning techniques to detect damage and diseases in crops, was presented in [12]. The system is structured into the following four subsystems: (1) smart irrigation system, (2) smart fertilizer-dose-recommendation system, (3) crop-disease-detection system, and (4) crop-damage-prediction system. Three convolutional neural network (CNN) architectures were implemented for disease prediction using multiclass image classification: ResNet50, VGG16 and DenseNet121. Regarding damage prediction, five machine-learning algorithms were used: LightGBM, XGBoost, random forest, decision tree and k-nearest neighbors (KNN), obtaining better results with the LightGBM algorithm.

Adel et al. [13] presented an architecture of an IoT-based smart monitoring system for agriculture, which was intended to give advice to farmers to avoid and prevent the spread of the late blight disease in potatoes and tomatoes. This system consists of three different layers, namely: (a) a perception layer consisting of data-acquisition nodes composed of sensors, microcontrollers and communication modules, (b) an application layer that displays all collected data to farmers through a dynamic web application and (c) a gateway layer that connects the perception layer with the application layer. Additionally, the authors implemented a prediction model using the linear regression technique and a classification model using a support vector machine (SVM) algorithm.

With the aim of significantly contributing to the saving of freshwater used in agriculture, especially in irrigation, an architecture of an intelligent autonomous irrigation system based on IoT technologies and machine-learning techniques was proposed in [14]. In particular, the system allows for predicting soil moisture using information collected from sensors deployed on the ground and weather forecast information extracted from the Internet through web services, thus helping to make effective irrigation decisions with optimal water use. A hybrid machine-learning algorithm based on a support vector regression algorithm and the k-means clustering algorithm was implemented for this purpose.

Li et al. [15] presented an intelligent agriculture system for the management and control of greenhouses. The system uses different IoT devices to collect a large amount of greenhouse environmental data and an improved k-means clustering algorithm based on the maximum distance method to select relatively optimal data as reference data for the next cycle in a greenhouse. It is structured into four major layers: (1) a sensors layer, which includes a variety of sensors, video cameras and other types of data-acquisition hardware, (2) a transport layer, which includes wireless communication and wired communication modules, (3) a business layer that is mainly responsible for the monitoring of the environmental data and (4) an application layer that allows interaction with the user through different web applications.

An expert system for the domain of agriculture, which is based on artificial intelligence techniques, specifically, on artificial neural networks, was proposed in [16]. This system helps farmers to assess land suitability for cultivation based on farming data obtained from an underlying wireless sensor network. In particular, these data are collected from different IoT-based sensors, including PH, soil moisture, salinity and electromagnetic sensors, using a Raspberry Pi Single-Board Computer (SBC), and it is then locally preprocessed and sent to the cloud to be stored for further processing. A multilayer perceptron (MLP)-based model for classification was finally implemented that exploits data stored in the cloud with the purpose of classifying land suitability for cultivation.

Alibabaei et al. [17] implemented recurrent neural network (RNN) models, namely, long short-term memory (LSTM), gated recurrent unit (GRU), bidirectional LSTM and bidirectional GRU models, to estimate tomato and potato yields at the end of a season based on time-series data, specifically, climate big data, irrigation scheduling data and soil water contents. Climate big data were collected by an agricultural weather station for a site in Portugal and retrieved from a government agency of the Ministries of Agriculture and the Sea. The performances of the models were compared with the performance of a convolutional neural network model, a multilayer perceptron model and a random forest regression model, and the results showed that the bidirectional LSTM model outperformed all alternative and baseline models in predicting tomato and potato yields.

A comparative analysis of the works described above is summarized in Table 1. This analysis comprises the following criteria of comparison: purpose, use of IoT technologies, use of data-mining techniques, use of machine-learning/artificial intelligence techniques, machine-learning task implemented, crop studied, use of crop-production data and use of climate data.

As is shown in Table 1, the work by Rezk et al. [10] has most of the features considered in the comparative analysis, which means that it is the work that is most similar to our work. Nonetheless, unlike our proposal, which aims to predict volume of production of crops, in their work a classification model is proposed to classify crop productivity and drought. Additionally, we focused on predicting the volume of corn grain production in the northeast region of the Mexican state of Puebla; corn grain is one of the most widely cultivated crops in the state of Puebla, Mexico, along with coffee beans and black beans. Conversely, Rezk et al. [10] focused on classifying productivity and drought of four different crops that are widely cultivated in the state of Maharashtra, India, namely, bajra, soybean, jowar and sugarcane. Furthermore, unlike most of the works analyzed, our work proposes a data-mining process that integrates two heterogeneous sources of crop-production data and climate data.

Table 1. Comparative analysis of related works.

Criterion/Work	[10]	[11]	[12]	[13]
IoT technologies	✔ (IoT sensor-based dataset collection)	✖	✖	✔ (IoT sensor-based dataset collection)
Data-mining techniques	✔	✖	✔	✖
Machine-learning/artificial intelligence techniques	✔ (PART algorithm)	✔ (Linear Regression, Single-Layer Perceptron)	✔ (Random Forest, KNN, LightGBM, RestNet50, VGG26 and DenseNet121)	✔ (SVM and Linear Regression)
Machine-learning task	Classification	Time-series forecasting	Classification	Prediction and classification
Crop studied	Bajra, soybean, jowar and sugarcane	Pear and apple	Tomato, potato, corn, apple and peach	Tomato and potato
Purpose	Drought and crop productivity	Crop harvest	Crop damage and disease	Crop disease
Crop-production data	✔	✖ (Soil productivity data)	✖ (Crop-fertilization data)	✖
Climate data	✔	✖	✖	✔
Criterion/Work	[14]	[15]	[16]	[17]
IoT technologies	✔ (IoT sensor-based soil moisture data collection)	✔ (IoT sensor-based dataset collection)	✔ (IoT sensor-based dataset collection)	✖
Data-mining techniques	✖	✔	✔	✔
Machine-learning techniques/artificial intelligence techniques	✔ (SVR and K-means)	✔ (K-means)	✔ (Multilayer Perceptron)	✔ (LSTM, GRU, bidirectional LSTM, bidirectional GRU)
Machine-learning task	Prediction and clustering	Clustering	Classification	Time-series forecasting
Crop studied		Unknown	Unknown	Tomato and potato
Purpose	Soil mosture	Greenhouse environmental factors optimization	Land suitability for cultivation	Crop yield
Crop production data	✖	✖	✖	✖
Climate data	✔ (Public data available on the Internet)	✖ (Greenhouse environmental data)	✖ (Soil productivity data)	✔

3. Smart Peasant Farming System and Data-Mining Process

The architectural design of the smart peasant farming system, which is the salient contribution of this work, is shown in Figure 2. The components of the proposed system architecture are described in the following subsections. Additionally, the data-mining process for crop-production prediction is described in the context of the description of the IoT data analytics component, which is one of the major components of the system architecture.

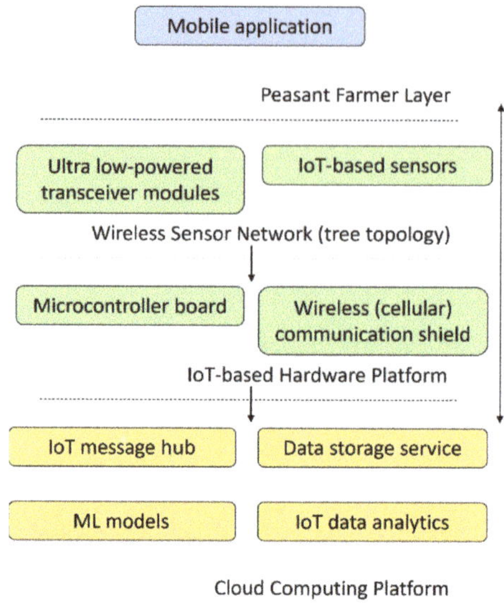

Figure 2. System Architecture.

3.1. *Peasant Farming Layer*

This layer basically consists of a mobile application designed for the peasant farmers to in-field register all of their crop yield data; data that are commonly collected and openly published by government agencies for statistical purposes, such as the area (in hectares) sown with a crop, the harvested area (in hectares) of a crop and the yield volume (in tons) of the harvested area of a crop.

Likewise, the mobile application is intended to show the peasant farmers the results of the data-mining process, i.e., the results of the crop yield prediction task on the integrated historical dataset: crop yield data + climate data. It is also intended to serve as a tool for the real-time monitoring of these data as it is remotely collected using a variety of in-field sensors. In this context, Internet of Things platforms such as Blynk, Ubidots and Arduino Cloud could be exploited to build web and mobile dashboards using user interface drag and drop editors, i.e., without programming any code.

3.2. *Wireless Sensor Network*

The Wireless Sensor Network (WSN) represents the second layer of the architecture of our system, which is based on ultralow powered transceiver modules designed for operation in the worldwide Industrial, Scientific and Medical (ISM) frequency band at 2.4 GHz such as the nRF24L01+ single chip 2.4 GHz transceiver (https://www.sparkfun.com/datasheets/Components/SMD/nRF24L01Pluss_Preliminary_Product_Specification_v1_0.pdf (accessed on 15 November 2021)), as well as on a variety of low-power sensors for climate and environmental monitoring. In this regard, the selection of the necessary sensors should favor low-cost sensors that are compatible with hardware and software prototyping platforms such as Arduino and NodeMCU, which are popular and relatively low-cost alternatives for the development of IoT-based systems.

The choice of this radio-frequency (RF) wireless communication technology over other similar technologies that are equally proprietary, such as Long Range (LoRa), lies in the possibility of exploiting communication protocols that are specially designed to enable high-power data transmission and reception at lower power consumptions. In fact, in the context of peasant and family farming, high bandwidths should be prioritized over long transmission ranges as constant up links for real-time data streams are frequently deployed over narrow geographic areas.

Moreover, a typical tree topology in which one of the nodes acts as a base node and the others are central hubs or actual sensor nodes is proposed for the design of the WSN.

Unlike other network topologies for WSNs, namely, cluster topology and flat topology, tree topology has been demonstrated to save slightly more energy in data-acquisition applications. On the other hand, tree topology can perform worse than cluster topology and chain topology in terms of scalability; similarly, it can perform worse than chain and flat topologies in terms of topology management (overhead) [18]. Nonetheless, a smart farming system such as the one that we pursue in this work, which is aimed at peasant and family farming, is theoretically less likely to suffer from these problems as it would be composed of no more than a few dozen sensors.

3.3. *IoT-Based Hardware Platform*

The foundation of this layer is a general purpose microcontroller-based development board wired to the transceiver module of the WSN acting as the root node. In this regard, some general purpose electronic prototyping platforms such as Arduino and the family of Discovery Boards offered by STMicroelectronics have microcontroller-based development boards especially designed for the IoT, such as Arduino 33 IoT and B-L475E-IOT01A, respectively.

These microcontroller boards usually facilitate integration with IoT platforms, cloud services or mobile and web development platforms such as Blynk, Amazon Web Services, or Google Firebase, respectively; nevertheless, they tend to be relatively more expensive than general purpose microcontroller boards, e.g., Arduino UNO.

Furthermore, one of the major components of the IoT-based hardware platform layer is the wireless networking module which must enable access to the Internet to communicate with the cloud (represented by the cloud-computing platform layer).

Due to the characteristics of the domain of application of our system, it will not normally be located near WiFi access points, so using Wireless Local Area Networks to connect it to the Internet would not be a viable option. Therefore, we have chosen to use cellular networks as an alternative networking technology in this regard.

3.4. *Cloud-Computing Platform*

The fourth layer of the architecture of our system is composed of four cloud computing services of four different categories: cloud messaging service, data storage service, machine-learning service and IoT data analytics service.

3.4.1. IoT Message Hub Service

In general terms, a cloud messaging service enables a channel for bidirectional communication between IoT devices/mobile applications and the cloud. Beyond the obvious need to create data streams for all the variables that are commonly monitored by a variety of sensors in a smart farming system, bidirectional communication is crucial for a data-mining-based smart farming system to be able to show back to its users the results of the data-mining process that is commonly carried out in the cloud.

3.4.2. Data Storage Service

A data storage cloud service typically consists of a NoSQL database (a nonrelational database), either an object database or a JavaScript Object Notation (JSON)-based database. The importance of this component within the proposed architecture lies in the possibility of storing all data collected by sensors composing the WSN in a secure and fully scalable manner.

In addition, some cloud-computing solutions from this category that are part of platforms aimed at developing serverless web and mobile applications such as Google Firebase and Amazon AWS Amplify allow multiple client devices to directly connect to databases through bidirectional channels. This enables real-time synchronization of data on all devices directly connected to a database in response to changes made on each of the devices, which would partially eliminate the need for a cloud messaging service. Nonetheless, there would be a need for a service that allows for sending messages from the cloud back to the connected devices because of the execution of functions hosted in the cloud, which would be the case of the crop yield prediction task in our system.

3.4.3. IoT Data Analytics Service

Regarding the data-mining process, all data stored in the NoSQL cloud database, which are raw data, must first be preprocessed to be transformed into data that can be used by machine-learning algorithms for the purpose of discovering knowledge. In particular, within the proposed architecture, this task is carried out by the IoT data analytics service.

Moreover, data preprocessing commonly involves data cleaning and data transformation itself. On the one hand, data cleaning means fixing or removing anomalies in data, and, in its simplest form, it is reduced to dealing with missing values, removing irrelevant values and removing duplicated values. This is crucial for an IoT sensor-based smart farming system because data captured from the physical world through sensors (in our case, climate data) tend to be noisy and unreliable.

On the other hand, data transformation typically involves data scaling and data normalization. Data scaling means fitting data within specific scales, whereas data normalization implies scaling data with the intention of transforming them to be normally distributed.

In this context, it is worth noting that popular cloud platforms (Platform as a Service, PaaS) such as Microsoft Azure and Amazon AWS include services that allow for automatizing common data preprocessing tasks, from data cleaning tasks through to data transformation tasks. On these cloud platforms, data can also be automatically preprocessed before being stored.

Figure 3 shows a flowchart representation of the data-mining process pipeline proposed in this work.

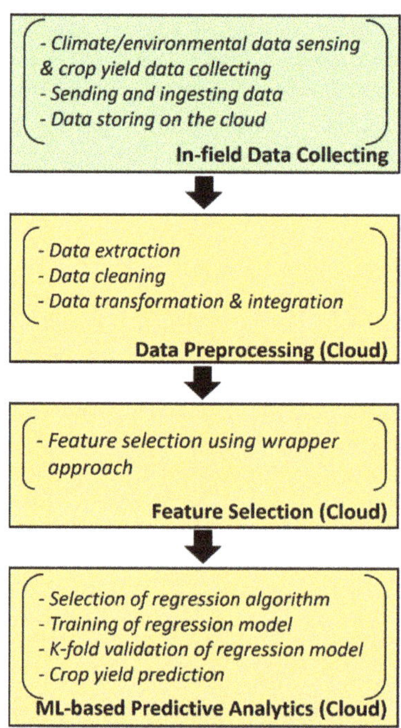

Figure 3. Data-Mining Process Pipeline.

3.4.4. Machine-Learning Service

One of the major components of the cloud-computing platform layer is the machine-learning service, which allows for implementing predictive algorithms and works in conjunction with the IoT data analytics service to enable machine-learning-based predictive analytics for the purpose of predicting crop yield.

In this context, being considered a de facto computational notebook for data science, Jupyter Notebook is supported by many of the most popular cloud platforms as a rapid, iterative and interactive way of implementing machine-learning algorithms for data mining, which mainly includes splitting datasets into training and validation sets as well as training and validating machine-learning models. In this work, we have chosen k-fold cross-validation as the preferred method for training and validating predictive models as it is the most recommended method for machine-learning model evaluation [19].

Implementing machine-learning algorithms also implies selecting those that are theoretically more appropriate given the nature of the data to be processed and the type of data-mining task to be carried out (in our case, prediction). In addition, it implies analyzing target data to identify those features or variables that are more relevant for use in generating machine-learning models, a task that is known as feature selection.

Feature selection can be carried out using two main different approaches: wrapper-based approaches and filter-based approaches. On the one hand, with the wrapper-based approaches, multiple machine-learning models are evaluated using procedures that incrementally add or remove features to find the approximately optimal combination that maximizes model performance. These procedures are mostly realized by greedy search algorithms; a greedy algorithm is any algorithm that follows the problem-solving heuristic of making the locally optimal choice at each stage.

On the other hand, filter-based approaches allow for evaluating the relevance of the features outside of the machine-learning models using statistical calculations, keeping only the features that pass some criterion. Unlike filter-based approaches, wrapper-based approaches completely depend on the underlying machine-learning algorithms and tend to be computationally intensive; they, however, usually provide the best-performing feature set for a particular type of machine-learning model [20].

This component is finally responsible for making crop yield predictions by exploiting resulting machine-learning models.

4. Materials and Methods

For a preliminary validation of the proposed architecture, a climate dataset was collected containing the following data observed during the period of 2003–2019 in the municipalities of Teziutlán, Tlatlauquitepec, Hueyapan, Hueytamalco, Yaonáhuac, Acateno, Atempan, Teteles de Ávila Castillo, Zaragoza, Chignautla, Ayotoxco de Guerrero, Zacapoaxtla, Cuetzalan de Progreso of the northeast region of the Mexican State of Puebla:

- Monthly average temperatures;
- Days with hail events per month;
- Days with fog occurrence per month;
- Days with storm activity per month;
- Total monthly rainfall.

Similarly, a crop yield dataset was collected that contains the following data observed during the same period in the same municipalities of the Mexican State of Puebla for both corn crops:

- Hectares planted with corn;
- Hectares harvested for corn;
- Production volumes (in tons) of the harvested areas of corn.

Figure 4 shows a map of the previously mentioned municipalities of the Mexican State of Puebla.

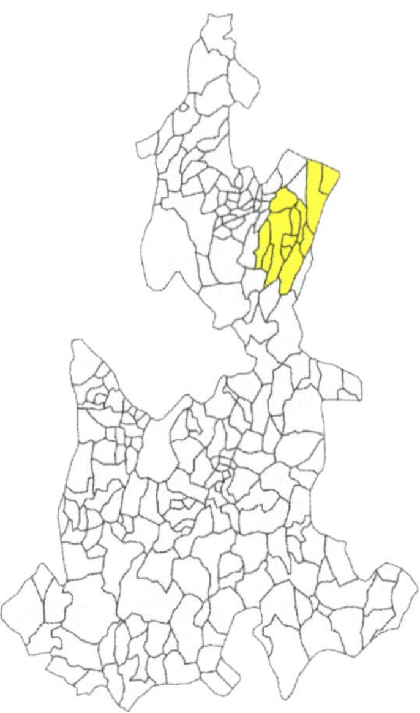

Figure 4. Selected municipalities of the Mexican State of Puebla.

These datasets were manually collected from the website of Mexico's National Water Commission (https://smn.conagua.gob.mx/es/informacion-climatologica-por-estado?estado=pue (accessed on 7 December 2021)) and the website of Mexico's Agri-food Information Service (http://infosiap.siap.gob.mx/gobmx/datosAbiertos.php (accessed on 7 December 2021)), respectively. The latter publishes annualized data on agricultural production at national, state and municipal levels as open data available in the form of spreadsheet files, whereas the former publishes a variety of data collected by the country's weather stations as annualized data by state in plain text format.

Table 2 summarizes the locations of the weather stations of the northeast region of the Mexican State of Puebla used as sources of climate data in this study.

Table 2. Weather stations.

Weather Station Name	Municipality	Location
"Teziutlán"	Teziutlán	Latitude: 19°49′49″ N. Longitude: 097°21′00″ W. Altitude: 1818.0 MASL
"Oyameles"	Tlatlauquitepec	Latitude: 19°42′51″ N. Longitude: 097°32′51″ W. Altitude: 2670.0 MASL
"Las Margaritas"	Hueytamalco	Latitude: 19°59′14″ N. Longitude: 097°17′14″ W. Altitude: 2422.0 MASL
"San José Acateno"	Acateno	Latitude: 20°08′24″ N. Longitude: 097°12′04″ W. Altitude: 144.0 MASL
"Zaragoza"	Zaragoza	Latitude: 19°47′10″ N. Longitude: 097°33′10″ W. Altitude: 2493.0 MASL
"Los Humeros (CFE)"	Chignautla	Latitude: 19°40′45″ N. Longitude: 097°24′22″ W. Altitude: 2862.0 MASL
"Ayotoxco de Guerrero"	Ayotoxco de Guerrero	Latitude: 20°05′43″ N. Longitude: 097°25′43″ W. Altitude: 237.0 MASL
"Zacapoaxtla (SMN)"	Zacapoaxtla	Latitude: 19°52′18″ N. Longitude: 097°35′18″ W. Altitude: 1828.0 MASL
"Cuetzalan de Progreso"	Cuetzalan de Progreso	Latitude: 20°02′20″ N. Longitude: 097°31′20″ W. Altitude: 756.0 MASL

As shown in Table 2, for most of the municipalities of our interest (9 out of 13), there is one local weather station providing data for the period of reference (2003–2019), which we used as the climate data source in this study.

Moreover, Table 3 summarizes the number of instances, features and missing values in each dataset. Notice that these statistics correspond to the unprocessed datasets. Details of handling missing data and reducing features (selecting features) are given in the following subsections.

Table 3. Unprocessed datasets' statistics.

Dataset	Number of Instances	Number of Features	Number of Missing Values
Climate dataset	221	62	1209
Crop yield dataset	442	6	0

The following section describes the variables included in the second dataset.

4.1. Crop Yield Dataset Description

Among the main variables included in the crop yield dataset are those described in Table 4. This dataset includes other variables used as reference: year, production cycle name and municipality name, which are not shown in Table 4. The year and municipality name variables are also included in the climate dataset.

Table 4. Variables in crop yield dataset.

Variable	Description	Unit of Measurement
Total cropped area	The total area planted with a crop	Hectares
Total harvested area	The total area harvested for a crop	Hectares
Production volume	The harvested production of a crop	Tons

4.2. System Construction

We partially implemented the proposed architecture to preliminarily validate it. First, we used NodeMCU as the microcontroller-based development board and the SIM 900 GSM/GPRS shield as the wireless networking module for the realization of the IoT-based hardware platform layer.

Second, we used Amazon's AWS IoT Core, AWS IoT Analytics and S3 services to realize the IoT message hub, IoT data analytics and data storage service components of the cloud-computing platform layer of the proposed architecture. Likewise, we used Jupyter Notebook documents to implement machine-learning algorithms for predictive analytics in Python, being able to run the resulting models on AWS IoT analytics thanks to the integration of this cloud service with the Jupyter Notebook data science tool.

4.3. Data Preprocessing on the Cloud

The datasets collected were ingested into Amazon S3 buckets, then these data were sent from the Amazon S3 service to the AWS IoT analytics service for data preprocessing purposes.

In particular, the climate and crop yield datasets were cleaned and transformed separately and then integrated into a single dataset. The integrated dataset comprised approximately 400 samples or observations.

Regarding data cleaning, the crop yield dataset required minimum treatment. On the contrary, the climate dataset required deeper treatment due to missing data (see Table 3). Any missing value of the variables representing monthly average temperatures, total monthly rainfalls, days with hail events in a month, days with fog occurrence in a month and days with storm activity in a month were calculated as the average of all the values registered for the corresponding month for all the years included in the dataset. In this context, notice that there are novel proposals for data-mining methods that inherently deal with missing values [21].

Regarding data transformation, average temperatures per production cycle (spring–summer and autumn–winter production cycles), as well as total rainfall per production cycle, days with hail events per production cycle, days with fog activity per production cycle and days with storm activity per production cycle were accordingly calculated from the variables representing monthly average temperatures, total monthly rainfalls, days with hail events in a month, days with fog occurrence in a month and days with storm activity in a month in the case of the climate dataset. As a result, the climate dataset was restructured as shown in Table 5.

Table 5. Variables in climate dataset.

Variable	Description	Unit of Measurement
Average temperature_spring	Average temperature during spring–summer production cycle.	Degrees Celsius
Average temperature_autumn	Average temperature during autumn–winter production cycle.	
Days with hail activity_spring	Days with hail activity during spring–summer production cycle.	Days
Days with hail activity_autumn	Days with hail activity during autumn–winter production cycle.	
Days with fog occurrence_spring	Days with fog occurrence during spring–summer production cycle.	
Days with fog occurrence_autumn	Days with fog occurrence during autumn–winter production cycle.	
Days with storm activity_spring	Days with storm activity during spring–summer production cycle.	
Days with storm activity_autumn	Days with storm activity during autumn–winter production cycle.	
Total rainfall_spring	Total rainfall during spring–summer production cycle.	Millimeters
Total rainfall_autumn	Total rainfall during autumn–winter production cycle.	

For the integration of the crop yield and climate datasets into a single dataset, we implemented a database-style joining approach by which new observations were generated by joining the observations from the latter with those from the former using the year and municipality name variables as indexes (see Figure 5). The rationale behind this is that crop production data must be interpreted in the context of the data about the weather conditions of the cropping areas.

From this perspective, the importance of the climate data in this work lies in the use that we made of it to enrich the crop yield data.

In addition, categorical variables in the integrated dataset, namely, production cycle name and municipality name were transformed into numerical variables using dummy variables.

Finally, the range of data in the integrated dataset was rescaled into a [0, 1] range, i.e., it was normalized.

4.4. Feature Selection and ML-Based Predictive Analytics on the Cloud

For feature selection, we preferred a wrapper-based approach to a filter-based approach. Because of this choice, we performed the feature selection step in conjunction with the machine-learning-based data analytics step of the proposed data-mining method.

In particular, we implemented the recursive feature elimination (RFE) method [22] to perform feature selection on the integrated dataset. RFE is an instance of the backward feature elimination method, which consists of an iterative process that implies training a classifier or regressor, computing the ranking criterion for all features involved and removing the feature with the smallest ranking criterion.

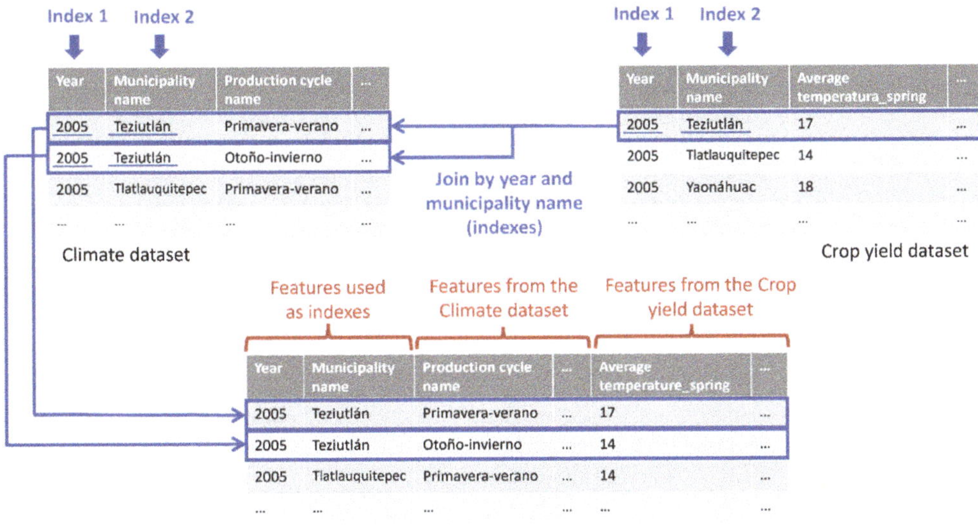

Figure 5. Database-style joining approach for dataset integration.

Furthermore, we used the linear regression and k-nearest neighbors (KNN) regression machine-learning algorithms in the core of the RFE algorithm as we selected them for the implementation of machine-learning-based predictive analytics in this work. In regard to the latter, we experimentally set the number of neighbors (k) at 5.

We carried out the following iterative procedure to select the optimal *k* value:

1. Initialize a random *k* value between 2 and N-1 (where N is the number of samples in our dataset);
2. Train the prediction model using the selected *k* value. Use the trained model to make predictions for the data in the dataset reserved for validation;
3. Calculate the Root-Mean-Square Error (RMSE) for the predictions computed;
4. Repeat the process selecting a different *k* value.

We finally created a plot between the RMSE values and the *k* values to select the *k* value with the minimum error rate, which was 5 ($k = 5$).

Linear regression and KNN regression are, respectively, easy and simple parametric and nonparametric machine-learning algorithms [23,24], and they were judged as theoretically appropriate for the nature of both the problem that we faced and the data that were available.

We sought to significantly reduce the number of features to employ to build prediction models using these algorithms; therefore, we experimentally set the threshold of relevant features at nine by carrying out the following iterative procedure:

1. Initialize a random value between 2 and F-1 (where F is the total number of features in our dataset) for the number of features to be selected (*f*);
2. Perform recursive feature elimination (RFE) to select the *f* most relevant features from the set of features in our dataset;
3. Train the prediction model using the selected *f* most relevant features. Use the trained model to make predictions for the data in the dataset reserved for validation;
4. Calculate the Root-Mean-Square Error (RMSE) for the predictions computed;
5. Repeat the process selecting a different *f* value.

We finally created a plot between the RMSE values and the *f* values to graphically select the f value with the minimum error rate, which was 9 ($f = 9$).

For both prediction models, the procedure resulted in the selection of the following nine features from our integrated dataset: total cropped area, total harvested area, year, total rainfall_spring, total rainfall_autumn, average temperature_spring, average temperature_autumn, days with hail activity_spring and days with hail activity_autumn. In particular, we trained and validated the prediction models using the k-fold cross-validation method, for which we set the number of folds (k) at 5 [19]. As a result, the integrated dataset was split into five consecutive folds, from which one was used once for the test while the four remaining folds were used once to train the models.

5. Results

Table 6 shows the results of the calculation of the default performance metric for prediction models in the scikit-learn library, in this case, for both the linear regression model and the k-nearest neighbors regression model, namely, coefficient of determination (R^2) [25]. This metric provides a measure of how likely the model is to predict unseen samples through the proportion of variance that is explained by the independent variables in the model [26]. In particular, this table shows the means of the R^2 scores computed for both prediction models in each step of the 5-fold cross-validation, as well as the standard deviations of these R^2 scores.

Table 6. Coefficient of Determination (R^2) scores.

Model	Coefficient of Determination (R^2) Scores	
	Mean	**Standard Deviation**
Linear Regression Model	0.756	+/−0.005
KNN Regression Model	0.944	+/−0.001

We must be careful in judging the high mean R^2 scores, because according to some statistical tests that we carried out there is collinearity between some of the predictor variables in our integrated dataset. We decided not to address this issue in this study because collinearity does not tend to influence the ability of a prediction model to predict new observations [26,27] and the goal of this study was to make accurate predictions. Nevertheless, in future work, we need to analyze to what extent these high R^2 scores are an indication of the collinearity problems existing in our integrated dataset.

Additionally, the standard deviations of the R^2 scores suggest that there is a very little variation in the performance of our prediction models when using different subsets of training data.

Furthermore, to support the R^2 scores obtained, we created estimated-by-observed plots for the learned linear regression model and the learned KNN regression model. Furthermore, we performed a graphical residual analysis on the results of the models to assess the assumptions of the regression models. In particular, we created a residuals vs. fits plot and a histogram of residuals for each of the models. Notice that a residual represents the vertical distance between an observed data point and its estimated value.

Figure 6 shows the estimated-by-observed plots, whereas the residuals vs. fits plots are shown in Figure 7 and the histograms of residuals are shown in Figure 8.

As shown in Figure 6, the estimated values of the KNN Regression Model are more strongly correlated with the observed values than the estimated values of the Linear Regression Model. This is a clear indication of how accurate each model is with respect to the other, and it supports the R^2 scores obtained, which can be interpreted as follows for the linear regression model and the KNN regression model, respectively:

- In total, 75.6% of the variation in response y (production volume) is accounted for by the variation in the set of predictors X;
- In total, 94.4% of the variation in response y (production volume) is accounted for by the variation in the set of predictors X.

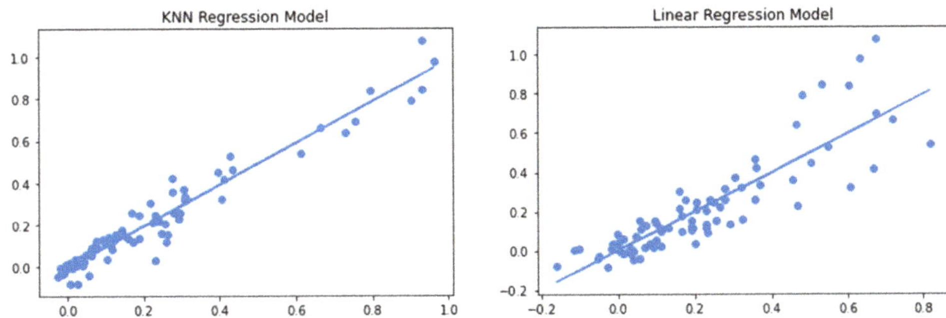

Figure 6. Predicted-by-observed plots.

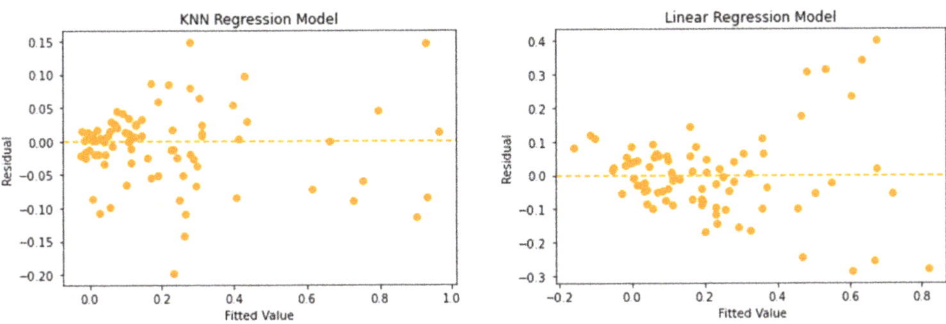

Figure 7. Residuals vs. fits plots.

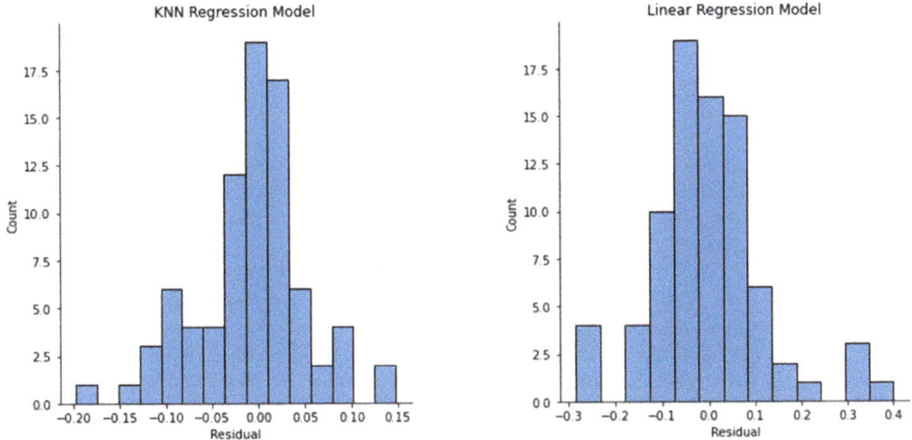

Figure 8. Histograms of residuals.

Recall that the set of predictors X resulting from the feature-selection process comprised the following predictors (features): total cropped area, total harvested area, year, total rainfall_spring, total rainfall_autumn, average temperature_spring, average temperature_autumn, days with hail activity_spring and days with hail activity_autumn.

Furthermore, as shown in Figure 7, the residuals of both the linear regression model and the KNN regression model are randomly scattered around the residual = 0 line, which indicates that the assumption of linear relationship is reasonable; nonetheless, the average of the residuals remains closer to 0 in the case of the KNN regression model than in the case of the linear regression model.

In addition, as shown in this figure, the variation of the residuals appears to be roughly constant at every level of the fitted values for the KNN regression model, which is evidence that the assumption of constant variance is not violated. The residuals vs. fits plot of the linear regression model shows approximately eight data points that can be judged as outliers as they do not follow the general trend of the rest of the data. Therefore, we should be careful in stating that the assumption of constant variances is also not violated in the case of the linear regression model; we should perform tests of equality of variances to check this in future work.

Finally, the bell-shaped appearance of the histograms of the residuals of the linear regression model and the KNN regression model, which are shown in Figure 8, is a clear indication that the assumption of normality is reasonable.

Moreover, we computed error rates for our final prediction models. In particular, we selected the Root-Mean-Square Error (RMSE) metric, which is a risk metric that corresponds to the expected value of the square root of the quadratic error or loss (see Table 7). RMSE is a very commonly used general purpose metric for numerical predictions [28].

Table 7. Root-Mean-Square Error (RMSE) rates.

Model	Root-Mean-Square Error (RMSE) Rate
Linear Regression Model	0.122
KNN Regression Model	0.058

Unlike R^2 scores, which should be close to 1, RMSE rates should be close to 0. As shown in Table 7, the RMSE rates for the linear regression model and the KNN regression model are 0.122 and 0.058, respectively, which are certainly close to 0. In addition, the RMSE rate of the KNN regression model is closer to 0 than the RMSE rate of the linear regression model. This finding is in correspondence with the finding of the previous performance analysis, which was carried out based on the R^2 metric, and it represents a strong indication of how well each prediction model performs with respect to the others.

6. Discussion

Judging by the results of the evaluation performed, it is feasible to use data-mining techniques to integrate heterogeneous crop production and climate data and to exploit it using traditional machine-learning techniques to reliably predict the volume of production of crops (t).

One of the major challenges that we faced in the integration of crop-production data and climate data was the difference in data granularity: crop-production data were available as annualized data whereas climate data were available as monthly data. This led us to perform simple temporal aggregation on the climate data, which intuitively implies the loss of some information as the number of observations is reduced. Notice that this information loss could naturally influence prediction performance negatively.

A linear relationship between the production volume variable and other crop production and climate variables such as total cropped area (t), total harvested area (t), average temperature (°C) and total rainfall (mm) was proven to exist using corn grain production data and climate data from the northeast region of the Mexican state of Puebla.

Crop-production predictions can be considered a starting point for the farmers to be able to make effective decisions in a timely manner based on reliable findings made before harvesting and even just after planting at the beginning of a production cycle. In fact, the potential benefits that crop-production predictions could have in planning crop production cycles should be assessed in future work. It is clear, however, that for the peasant farming families in the northeast region of the State of Puebla, Mexico, the possibility of accessing low-cost technology that allows them to maximize crop production is of great importance, considering that, in many cases, agricultural production directly represents their primary source of food.

In future work, a smart peasant farming system should also be constructed based on the architecture proposed in this work to be able to compile a dataset from crop production and climate data collected using the set of in-field IoT-based sensors and the mobile application conceived for that purpose. This will allow us to carry out a validation of our proposal as a whole. Alternatively, regarding climate data, we should explore the use of satellite imagery as a possible data source. In fact, the other big challenge that we faced in this study was the shortage of climate data due to the scarcity of local weather stations across the selected municipalities of the Mexican State of Puebla.

Additionally, we will study the suitability of using other traditional supervised machine-learning algorithms and artificial intelligence techniques such as artificial neural networks to build more accurate prediction models. In this context, we plan to explore a hybrid supervised/unsupervised machine-learning approach in which clustering algorithms (unsupervised machine learning) are used to automatically label our data and machine-learning algorithms for prediction are used to make predictions based on the labeled data [29,30].

Regarding feature selection, we will study the feasibility of solving our prediction problem using regression techniques that directly reduce the set of predictive variables to the smaller set of uncorrelated variables, such as partial least squares (PLS) regression and principal component regression.

Furthermore, we will address the problem of predicting the volume of production of other crops that are popular in the northeast region of the state of Puebla, Mexico, such as coffee beans and black beans, including data that allow for capturing climate variability such as the minimum and maximum temperatures.

Finally, we plan to study the feasibility of integrating semantic web technologies for knowledge representation and reasoning, and the integration of recommendation techniques based on these technologies to the proposed architecture to improve the data-mining process [31,32].

Author Contributions: All authors contributed equally to the work. All authors have read and agreed to the published version of the manuscript.

Funding: This research was funded by Tecnológico Nacional de México, grant number 337533 (11079). This paper is also part of the research project LaTe4PSP (PID2019-107652RB-I00) funded by MCIN/AEI/10.13039/501100011033.

Institutional Review Board Statement: Not applicable.

Informed Consent Statement: Not applicable.

Data Availability Statement: The data presented in this study are available on request from the corresponding author. The data are not publicly available due to privacy restrictions.

Conflicts of Interest: The authors declare no conflict of interest.

References

1. Witten, I.H.; Frank, E.; Hall, M.A.; Pal, C.J. Chapter 1—What's it all about? In *Data Mining*, 4th ed.; Witten, I.H., Frank, E., Hall, M.A., Pal, C.J., Eds.; Morgan Kaufmann: Burlington, MA, USA, 2017; pp. 3–41. [CrossRef]
2. Mohammed, J.Z.; Meira, W., Jr. *Data Mining and Machine Learning: Fundamental Concepts and Algorithms*, 2nd ed.; Cambridge University Press: Cambridge, UK, 2020.
3. Delen, D. *Predictive Analytics: Data Mining, Machine Learning and Data Science for Practitioners*, 2nd ed.; Pearson FT Press: Hoboken, NJ, USA, 2020.

4. Dlodlo, N.; Kalezhi, J. The internet of things in agriculture for sustainable rural development. In Proceedings of the 2015 International Conference on Emerging Trends in Networks and Computer Communications (ETNCC), Windhoek, Namibia, 17–20 May 2015; pp. 13–18. [CrossRef]
5. Tzounis, A.; Katsoulas, N.; Bartzanas, T.; Kittas, C. Internet of Things in agriculture, recent advances and future challenges. *Biosyst. Eng.* **2017**, *164*, 31–48. [CrossRef]
6. Ray, P.P. Internet of things for smart agriculture: Technologies, practices and future direction. *J. Ambient. Intell. Smart Environ.* **2017**, *9*, 395–420. [CrossRef]
7. Shi, X.; An, X.; Zhao, Q.; Liu, H.; Xia, L.; Sun, X.; Guo, Y. State-of-the-Art Internet of Things in Protected Agriculture. *Sensors* **2019**, *19*, 1833. [CrossRef] [PubMed]
8. Colizzi, L.; Caivano, D.; Ardito, C.; Desolda, G.; Castrignanò, A.; Matera, M.; Khosla, R.; Moshou, D.; Hou, K.-M.; Pinet, F.; et al. Chapter 1–Introduction to agricultural IoT. In *Agricultural Internet of Things and Decision Support for Precision Smart Farming*; Castrignanò, A., Buttafuoco, G., Khosla, R., Mouazen, A.M., Moshou, D., Naud, O., Eds.; Academic Press: Cambridge, MA, USA, 2020; pp. 1–33. [CrossRef]
9. He, Y.; Zhang, Q.; Nie, P. Introduction of Agricultural IoT. In *Agricultural Internet of Things: Technologies and Applications*; He, Y., Nie, P., Zhang, Q., Liu, F., Eds.; Springer International Publishing: Cham, Switzerland, 2021; pp. 1–21. [CrossRef]
10. Rezk, N.G.; Hemdan, E.E.-D.; Attia, A.-F.; El-Sayed, A.; El-Rashidy, M.A. An efficient IoT based smart farming system using machine learning algorithms. *Multimed. Tools Appl.* **2021**, *80*, 773–797. [CrossRef]
11. Balducci, F.; Impedovo, D.; Pirlo, G. Machine Learning Applications on Agricultural Datasets for Smart Farm Enhancement. *Machines* **2018**, *6*, 38. [CrossRef]
12. Garg, S.; Pundir, P.; Jindal, H.; Saini, H.; Garg, S. Towards a Multimodal System for Precision Agriculture using IoT and Machine Learning. In Proceedings of the 2021 12th International Conference on Computing Communication and Networking Technologies (ICCCNT), Kharagpur, India, 6 June 2021; pp. 1–7.
13. Araby, A.A.; Elhameed, M.M.A.; Magdy, N.M.; Said, L.A.; Abdelaal, N.; Allah, Y.T.A.; Darweesh, M.S.; Fahim, M.A.; Mostafa, H. Smart IoT Monitoring System for Agriculture with Predictive Analysis. In Proceedings of the 2019 8th International Conference on Modern Circuits and Systems Technologies (MOCAST), Thessaloniki, Greece, 13–15 May 2019; pp. 1–4. [CrossRef]
14. Goap, A.; Sharma, D.; Shukla, A.K.; Krishna, C.R. An IoT based smart irrigation management system using Machine learning and open source technologies. *Comput. Electron. Agric.* **2018**, *155*, 41–49. [CrossRef]
15. Li, C.; Niu, B. Design of smart agriculture based on big data and Internet of things. *Int. J. Distrib. Sens. Netw.* **2020**, *16*, 1550147720917065. [CrossRef]
16. Vincent, D.R.; Deepa, N.; Elavarasan, D.; Srinivasan, K.; Chauhdary, S.H.; Iwendi, C. Sensors Driven AI-Based Agriculture Recommendation Model for Assessing Land Suitability. *Sensors* **2019**, *19*, 3667. [CrossRef] [PubMed]
17. Alibabaei, K.; Gaspar, P.D.; Lima, T.M. Crop Yield Estimation Using Deep Learning Based on Climate Big Data and Irrigation Scheduling. *Energies* **2021**, *14*, 3004. [CrossRef]
18. Mamun, Q. A Qualitative Comparison of Different Logical Topologies for Wireless Sensor Networks. *Sensors* **2012**, *12*, 14887–14913. [CrossRef] [PubMed]
19. Ozdemir, S. 12. Beyond the Essentials. In *Principles of Data Science: Learn the Techniques and Math You Need to Start Making Sense of Your Data: Mathematical Techniques and Theory to Succeed in Data-Driven Industries*, 1st ed.; Packt Publishing: Birmingham, UK, 2016.
20. Mafarja, M.M.; Mirjalili, S. Hybrid binary ant lion optimizer with rough set and approximate entropy reducts for feature selection. *Soft Comput.* **2019**, *23*, 6249–6265. [CrossRef]
21. Dinh, D.-T.; Huynh, V.-N.; Sriboonchitta, S. Clustering mixed numerical and categorical data with missing values. *Inf. Sci.* **2021**, *571*, 418–442. [CrossRef]
22. Guyon, I.; Weston, J.; Barnhill, S.; Vapnik, V. Gene Selection for Cancer Classification using Support Vector Machines. *Mach. Learn.* **2002**, *46*, 389–422. [CrossRef]
23. Boehmke, B.; Greenwell, B. Chapter 4 Linear Regression. In *A Machine Learning Algorithmic Deep Dive Using R*, 1st ed.; Chapman and Hall/CRC: Boca Ratón, FL, USA, 2020.
24. Bartosik, A.; Whittingham, H. Chapter 7—Evaluating safety and toxicity. In *The Era of Artificial Intelligence, Machine Learning, and Data Science in the Pharmaceutical Industry*; Ashenden, S.K., Ed.; Academic Press: Cambridge, MA, USA, 2021; pp. 119–137. [CrossRef]
25. Glantz, S.; Slinker, B.; Neilands, T. Chapter Three: Regression with Two or More Independent Variables. In *Primer of Applied Regression & Analysis of Variance*, 3rd ed.; McGraw Hill/Medical: New York, NY, USA, 2016.
26. Kutner, M.; Nachtsheim, C.; Neter, J.; Li, W. Chapter 2 Inferences in Regression and Correlation Analysis. In *Applied Linear Statistical Models*, 5th ed.; McGraw-Hill/Irwin: New York, NY, USA, 2004.
27. Frost, J. Chapter 9—Checking Asssumptions and Fixing Problems. In *Regression Analysis: An Intuitive Guide for Using and Interpreting Linear Models*, 1st ed.; Statistics by Jim Publishing: Costa Mesa, CA, USA, 2020.
28. Neill, S.P.; Hashemi, M.R. Chapter 8—Ocean Modelling for Resource Characterization. In *Fundamentals of Ocean Renewable Energy*; Neill, S.P., Hashemi, M.R., Eds.; Academic Press: Cambridge, MA, USA, 2018; pp. 193–235. [CrossRef]
29. Dinh, D.-T.; Fujinami, T.; Huynh, V.-N. Estimating the Optimal Number of Clusters in Categorical Data Clustering by Silhouette Coefficient. In *Knowledge and Systems Sciences*; Springer: Singapore, 2019; pp. 1–17. [CrossRef]

30. Astolfi, D.; Pandit, R. Multivariate Wind Turbine Power Curve Model Based on Data Clustering and Polynomial LASSO Regression. *Appl. Sci.* **2022**, *12*, 72. [CrossRef]
31. Valencia-García, R.; Ruiz-Sánchez, J.M.; Vivancos-Vicente, P.J.; Fernández-Breis, J.T.; Martínez-Béjar, R. An incremental approach for discovering medical knowledge from texts. *Expert Syst. Appl.* **2004**, *26*, 291–299. [CrossRef]
32. García-Sánchez, F.; Colomo-Palacios, R.; Valencia-García, R. A social-semantic recommender system for advertisements. *Inf. Processing Manag.* **2020**, *57*, 102153. [CrossRef]

MDPI
St. Alban-Anlage 66
4052 Basel
Switzerland
Tel. +41 61 683 77 34
Fax +41 61 302 89 18
www.mdpi.com

Applied Sciences Editorial Office
E-mail: applsci@mdpi.com
www.mdpi.com/journal/applsci

www.ingramcontent.com/pod-product-compliance
Lightning Source LLC
LaVergne TN
LVHW070851170726
843515LV00005B/619